Revolutionary Biology

Revolutionary Biology

The New, Gene-Centered View of Life

David P. Barash

Transaction Publishers

New Brunswick (U.S.A.) and London (U.K.)

Second paperback printing 2004
Copyright © 2001 by Transaction Publishers, New Brunswick, New Jersey.

This book is printed on acid-free paper that meets the American National Standard for Permanence of Paper for Printed Library Materials.

Library of Congress Catalog Number: 2001017123
ISBN: 0-7658-0067-5 (cloth); 0-7658-0963-X (paper)
Printed in the United States of America

Library of Congress Cataloging-in-Publication Data

Barash, David P.
    Revolutionary biology : the new, gene-centered view of life / David P.
Barash
        p. cm.
    Includes bibliographical references and index.
    ISBN 0-7658-0963-X (alk. paper)
    1. Human genetics. 2. Human evolution.   3. Genetic psychology.
    I. Title.

QH431 .B226 2001
599.93' 5—dc21   2001017123

Dedicated to the memory of
William D. Hamilton (1936–2000)
without whom the revolution might not have happened

Contents

1

Revolutionary Biology: The Family Face

There is a revolution under way in biology. It is giving rise to a new vision of life in general, of human beings in particular, and a new way of understanding why people behave as they do. Fortunately, this revolution is altogether nonviolent, shedding light instead of blood. Like most revolutions, however, this one has generated a fair amount of soul-searching, as its participants find themselves looking afresh at things that had been taken for granted. (And like most revolutions, this one has also been bitterly resisted by many of those who remain committed to the old ways.)

The biological revolutionaries have begun to confront some fundamental questions that are important to us all. Such as: Why have children? Why love them? Why do fathers and mothers frequently disagree, not only about their own lives but even about their children? Why do parents and children also disagree so frequently? Why do children seem to want more than parents can give, and why do parents expect so much from their children? What about sibling rivalry? The Oedipus complex? Why is it that people can adopt and love children born to others, and yet, step-families are so difficult? Why care about relatives? Why treat them differently from strangers? Are people, or animals, naturally altruistic? Or is it normal to be selfish? Why do people love their country? Their football team? Can they act according to the Golden Rule? Should they?

More generally, are human beings basically altruistic and good, or selfish and bad? Why do people show such a complex and often troubling array of social and antisocial tendencies? Why, given that families are so important, are they often so difficult? Why are there

1

fewer saints than sociopaths? Even "Why are we here?" and—believe it or not—"What is the meaning of life?"

Thanks to this biological revolution, it is also possible to see the rest of nature in a new, provocative, and useful way, while also acknowledging human beings to be on the one hand, organic creatures like any other, and yet, at the same time, special. Thanks to this biological revolution, we can, at last, begin to make sense of families, and in the process, make sense of ourselves.

The basis of this revolution is a new paradigm with an old pedigree, a grand theory of life—in fact, *the* grand theory of life—with scope as broad as its agenda: evolution. Right at the outset, you might hear some people grumbling that evolution is "only a theory." Don't believe it. Or rather, don't be bamboozled. In particular, don't mistake "theory" for "hypothesis." In science, a hypothesis is a conjecture, an educated hunch that may or may not be true, and that is tested by research. A scientific theory, on the other hand, is a coherent body of principles tested many times over and used to explain a class of phenomena. Hypotheses, like informed guesses in general, come and go. Theories also come and go, but when—as in the case of evolutionary theory—they stick around, it is because they continue to be supported, besting their alternatives, until eventually they become the very substance of scientific knowledge.

"The theory of evolution" has weathered test after test, emerging from each challenge (and there have been many!) refined and strengthened. Thus, evolutionary theory is to biology what atomic theory is to chemistry, or number theory to mathematics, or gravitational theory to astronomy: as close to fact as science is likely to get. Evolutionary "theory" is the fundamental, unifying prism that focuses light on the meaning, structure, history, and dynamics of life on earth. This focus is intense, and can be inflammatory. But all recent developments in biological science and medicine are based on evolutionary theory, from advances in molecular biology to new understandings of infectious disease and progress in cancer prevention and treatment. As Charles Darwin first described it and subsequent researchers have filled in many of the blanks, evolution is a Rosetta Stone, the key to life on this planet—and, for all we know, wherever else it may appear.

It must be noted that in some ways this "revolution" is not all that new; in fact, it is becoming middle-aged. Following its Darwinian manifesto (nearly 150 years ago), even the major uprisings in revolutionary biology are more than twenty years old. But most scientific revolutions are not like the Russian Revolution or the Boston Tea Party, sudden conflagrations that are over quickly, even as their effects linger. Unlike a military or political coup that "only" involves a changing of the guard, scientific revolutions change something far more resistant: people's minds. And so, most scientific revolutions take decades, and although they rarely involve violence, sometimes they do not succeed until the old guard passes away.

In the evolutionary revolution—at least when applied to human behavior—the battle is far from over. As a matter of fact, mainstream opinion in the social sciences (psychology, anthropology, sociology, and so forth) still clings to the dogma that human behavior is overwhelmingly a product of culture, learning, and social tradition. When most social scientists mention evolution at all, it is often misconstrued; to paraphrase Winston Churchill, never have so many said so much about something they understood so little.

It is sometimes said that one way to tell an optimist from a pessimist is to give the subject a shovel and put him or her in a room filled with horse manure. While the pessimist does nothing, the optimist digs vigorously, reasoning that with so much manure, there must be a horse nearby! Evolution and its misconceptions have generated many rooms-full of horse manure, but despite all the foolishness, revolutionary biologists are optimists. (Be warned: shoveling ahead!)

While some people still believe in various creation myths, from the Judeo-Christian story of Adam and Eve to the Hindu principle of reincarnation, most scientists take evolution for granted, not because it is unquestioned, but exactly the opposite: throughout decades of searching questions, evolution has continued to provide answers, thereby proving itself, over and over.

Scientific confidence in the "correctness" of evolution is all right so far as it goes, but it doesn't go far enough. In particular, to see evolution as simply a historical process is to give it too little credit.

In the public mind, evolution is widely associated with dinosaurs, "cave-men," dusty museum exhibits, and maybe something about embryos having gills and tails. But—especially as it is understood by the modern revolutionaries known as evolutionary psychologists, sociobiologists, or behavioral ecologists—evolution is much more immediate and its implications, downright dramatic. It isn't dead, like the dinosaurs. It is alive and kicking, the driving force behind what people *are* and what they *do*, no less than where they came from. It is the dynamic stuff of human loves, hates, fears, and desires no less than of bones and other fossils. When we really understand evolution, then and only then will we really understand human beings.

This book, then, is about understanding ourselves, not so much how we came to be but *what* we have become, right now, today. We are not going to spend time on bones, internal chemicals, or brains, but rather on what makes human beings tick, how and why they act the way they do. That is where the imprint of evolution is most striking, and where modern revolutionary biology is most exciting.

Before we proceed, however, a major caveat is in order. Nothing, it seems, is as simple as most people might wish. Our minds crave simplicity; the universe responds with complexity. We look for either/or, yes/no answers; the world gives us subtlety and ambiguity—this *and* that, yin *and* yang, interpenetrating and inseparable. If this is true for such "simple" things as the nature of an electron (particle *and* wave), then consider how much more true it is for complex things such as the nature of human nature (biological *and* cultural, genetic *and* learned).

Nothing in this book should be taken as arguing that human beings are "only" DNA made manifest, or that human behavior can be entirely understood as the outcome of pushy genes, struggling to get ahead. Human nature also includes complex symbolic thought, social learning and cultural tradition, as well as—on occasion—plain old-fashioned stubborn refusal to go along with experience, sometimes a cantankerous insistence on defying laws and rules, including even those that are supposedly "biological." Human nature is just too fluid, too complicated, too difficult, to be captured entirely by any formulation, whether biological, philosophical, cultural, theological, or what-have-you. *Homo sapiens* is an ornery old bird, not readily captured in a single, simple net.

President Harry Truman once complained that he would give a lot to have a one-armed economist. It seems that whenever Truman sought advice, his economists would reply with their best guess, and then add "But on the other hand . . . " Because the insights of revolutionary biology are so new, and their promise is so great, this is pretty much a one-handed book. It is based on the premise that its brand of evolutionary thinking is worth pursuing, to see how far it will take us. But this does not mean that gene-centered evolutionary theory has cornered the market on what makes human beings tick, what makes them special, or even, what makes them interesting. There is always another hand.

For an analogy, imagine someone had just discovered, for the first time, that heart disease is influenced by diet. (Let's say that previously, everyone thought it was "purely genetic.") There would be great interest and enthusiasm in the new-found role of food in predisposing to heart disease. People would eagerly—and appropriately—investigate the details of saturated versus unsaturated fats, cholesterols of all sorts, the significance of proteins, minerals, overall calorie intake, and so forth. Books would be written, explaining the astounding new discovery that diet affects heart disease, even though the ghost of Harry Truman's economists could still be heard: "But on the other hand . . . there are additional causes of heart disease, those related to exercise, for example. Ditto for stress, genetic susceptibility factors, other pre-existing conditions such as rheumatic fever, etc."

As this book unfolds, please do not mistake its one-handed approach—arrogant and blinkered as it may seem—for a fundamentalist's certainty that there is only one avenue to truth. Rather, as we explore the implications, insights, and excitement of revolutionary, gene-centered biology, see it as an attempt to follow a novel and promising path to discover how far it will take us, fully aware that there are other paths, other truths, other factors that ultimately must be taken into account if anything approaching a valid "explanation" for human beings and their behavior is ever to be achieved. Occasionally—all too rarely, for those committed to other interpretations—we'll suggest alternative ways of explaining things, if only to restrain excessive enthusiasm for a biological approach which, although fruitful, is only one of many paths to understanding ourselves.

If it helps, keep in mind the story of the beloved rabbi who once listened, with great sympathy, as a man from his congregation complained bitterly about his wife. A few minutes later, the rabbi responded with equal sympathy to the wife's complaints about the husband, telling each separately and sympathetically, "I understand. You're absolutely right." (In late twentieth-century terms, he could "feel their pain.") Then, when both husband and wife were gone, the rabbi's wife, who had been listening all the while, berated him: "First you said the man was right, then you said the woman was right. This is impossible. They can't both be right!" The rabbi paused for a moment, and then replied calmly, "You know, you're right too."

The Gene in the Machine, or
The Invasion of the Body Snatchers Revisited

At the core of revolutionary biology is a new perception of the relationship of bodies and genes. In short, the basic idea is this: the chicken-and-egg problem is finally solved. The egg came first. Or, more to the point, the chicken is merely a servant of the egg, rather than the other way around. Samuel Butler was correct: the chicken is an egg's way of making more eggs.

Not that bodies aren't important. It is bodies that we see occupying the landscape, whether chicken bodies or human beings. Bodies, bodies everywhere. Bodies breathing, reproducing, fighting, sharing, arguing, running about or lying down, building buildings and writing novels, voting and emoting, making war and making love. Every person is carefully tucked into a body, occupying—or so we assume—neither more nor less than the exact boundaries of his or her own skin.

This may be a common-sense and even obvious perspective, but it can also be misleading. Consider, for example, that the best view in Warsaw, Poland, is from the top of the Ministry of Culture, a monument of Stalinesque architecture at its worst. Why, then, is the view so good? Because it is the only place in the city from which you cannot see the Ministry of Culture! Similarly, each of us looks out at the rest of the world from the private confines of our cozy little home and castle: our bodies, ourselves. The view is good, but not necessarily accurate.

It is very difficult for people to consider themselves as anything other than their bodies, taken as a whole. You may look, with detachment, at a sample of "your" red blood cells, wriggling on a microscope slide, at a dental x-ray, a chromosome enlargement, or an MRI of one part or another of your anatomy, all the while knowing that in some sense, it isn't really and truly "you."

From the perspective of evolution, who or what constitutes that individual you call yourself? And what does it mean to acknowledge that a lengthy process of natural selection has produced it? Taken at face value, insofar as every living thing is the product of evolution, it means that each individual "me" has somehow been squeezed and molded, pruned and pounded into shape by natural selection operating via evolution's impact on our ancestors.

Correct, but only up to a point. Thus, everyone is indeed the product of evolution by natural selection. Biology's revolution has shown, however, that the face who stares back from the mirror (just like the one who looks out at the world from that structural bastion we call a body) is only indirectly the handiwork of evolution. Rather, all bodies are the product of *genes*, and it is they—the genes—that are evolution's legacy to every person, made corporeal in all those bodies that people today's world. Your face, your body, your physical and biological self: all are the outer manifestations of your genes, after they have interacted with the food, the oxygen, the learning, the exercise, and the sum total of all the other experiences that eventually produced "you." It is human *genes*—and hummingbird genes, and hyacinth genes, and so on—that have been picked and pruned, some doing well and others poorly, tumbling over each other, some leaping ahead and others dying out, mutating on occasion or remaining utterly unchanged, flowing along as best they can in a continuously meandering stream of DNA for millions of years, beginning with the origin of life and with no end in sight.

Each individual, on the other hand, is something else again, more like a temporary eddy in that living, surging river. Look at it this way: "you" are only as old as the time that has passed since your birth. In a sense, it wasn't evolution that put you together so much as the laws of chance, which combined half of your mother's genes with half of your father's, then shook you gently with a hefty dose of experience and environment, then marinated the concoction in tincture of time.

In another sense, although chance determined which genetic cards you were dealt, it was evolution that produced those "cards" in the first place. The genes that carry the information needed for each cell and protein in your body were established in the eons that passed *before* you were born. Like playing cards, they have the potential to be reshuffled and reused, in different combinations, over and over again in a poker game that isn't merely "all night," but potentially for all time. You, by contrast, won't last nearly so long.

Each individual is analogous to a "hand" of bridge, pinochle, gin rummy, hearts, or poker (take your pick; there are many more potential people than potential arrangements of playing cards). Over an evening of card-playing, such "hands" are evanescent. It is the game (of life) that persists, and the cards (genes) that make it possible. At the same time, there are many ways of playing out the hand you are dealt. Your own face, its scars and lines and wrinkles, straight or crooked teeth, glasses or no glasses, no less than your sentiments, beliefs, memories, and dreams are unique to you and your personal developmental history, just as a hand of cards is unique to a given occasion. Sometimes, in the long road from potential to actual (biologists would say from "genotype" to "phenotype") your unique genetic legacy is hidden: although you may have been born with genes for crooked teeth, if you had orthodonture, your smile today reflects not only those genes but also a good dose of corrective—not to say expensive—experience.

But no matter how much money, or time, or effort is lavished on them, bodies don't have much of a future. In the scheme of things, they are as ephemeral as a spring day, a flower's petal, a gust of wind. What does persist? Not bodies, but *genes*. Bodies go the way of all flesh: ashes to ashes and dust to dust, molecule to molecule and atom to atom. Bodies are temporary. They are all put together from various recycled parts scavenged from the cosmic junk-heap. *Genes*, on the other hand, are potentially immortal.

In his poem "Heredity," Thomas Hardy had a premonition of modern evolutionary biology and the endurance of genes:

I am the family face
Flesh perishes, I live on
Projecting trait and trace
Through time to times anon,
And leaping from place to place
Over oblivion.
The years-heired feature that can
In curve and voice and eye
Despise the human span
Of durance—that is I;
The eternal thing in man,
That heeds no call to die.

When people refer to evolution with the verbal shorthand "survival of the fittest," only rarely do they understand just what they are talking about. The "fittest," as biologists use the term, are those who outreproduce their rivals; more precisely, those most successful in projecting copies *of their genes* into the future. (More precisely yet, and more in tune with revolutionary biology: the most fit genes are those that succeed in outdoing rival genes in promoting their success.) Fitness is what genes strive for; bodies are how they do it. Physical fitness often helps, but the bottom line isn't bodily strength, or even overall health. Rather, what matters to evolution is the ability to stay in the game, to survive and prosper over long stretches of time. The issue isn't health, happiness, prosperity, or even longevity (that is, a long-lived body) as such, since even a Methuselah will eventually fall by the evolutionary wayside, to be replaced by genetic lineages that somehow manage to prosper longer than the most elderly individual.

Survival of the fittest? Certainly. But zoologist Richard D. Alexander of the University of Michigan put it well when he emphasized that the crucial question is survival of the fittest *what*?[1]

It can't be the fittest individual body, since most individuals of most species are actually rather short-lived. In the billions of years that evolution has had to work with, natural selection has perfected all sorts of highly sophisticated, complex adaptations, absolute marvels of sophisticated design: mammals that fly and that use sonar to hunt moths on the wing, caterpillars that look like snakes and vice versa, bacteria that can prosper in superheated sea water, eagles that can make out the face of a dime while hovering a hundred feet in the air. Amid all these marvels, if there had been a significant evolutionary advantage for individuals to live for hundreds of years, it seems very likely that this could have been done, too. After all, in some cases it has. Redwood trees and bristlecone pines survive for literally *thousands* of years. And yet, these are notable exceptions. The answer—first pointed out by George C. Williams, one of the founders of revolutionary biology—is that mere survival of the individual is not very high on the evolutionary agenda; in fact the overwhelming majority of living things have a life span of less than a year.[2]

Natural selection is clearly capable of producing long-lived bodies, even among vertebrates, as witnessed by giant tortoises or sturgeon. Most of the time, however, it doesn't bother. In fact, scientists have discovered that there are genes for so called "apoptosis," meaning programmed cell death. These are genes that cause cells to commit suicide. Why? Because, once again, the survival of the individual isn't what evolution is "all about." Rather, it is survival of the "family face."

In short, survival of the genes.

Individual life-span is a useful example because it helps clarify the importance of genes versus bodies in the evolutionary balance sheet.

When long-enduring bodies are a good way to promote the genes in question, then long-enduring bodies will evolve. If living fast, loving hard, and dying young is the best strategy, then that is what evolution will produce. Best for what? Best for the genes, or rather, for success-fully projecting them into the future, and thus, for staying in the evolu-tionary card game. Once again: bodies are a means, not an end. If bodies instead of genes were the apple of evolution's eye, then almost certainly, those bodies would be a whole lot more impressive than they actually are. Instead, it is the genes that are "impressive," in that they, not the bodies they create, are capable of enduring for genera-tions. Theoretically, as Thomas Hardy pointed out, they "heed no call to die." Bodies aren't so fortunate.

It all makes sense, once we recognize that the individual exists for the benefit of the genes, not the other way around. Hence, a chicken really *is* an egg's way of making more eggs; in fact, from a strictly biological perspective, *it is nothing but that!* (This is a way of under-standing nature, not a value judgment or political statement on the importance of individual lives, human or otherwise. By analogy, we can say that the planet Earth is only a tiny ball of water, mud and metal in a vast universe. But it is *our* earth, *our* home, and very important to us, just as each individual, body, mind, and spirit is likely to be critically important to each human being.)

Sigmund Freud, of all people, prefigured a gene-centered perspec-tive, emphasizing—not surprisingly—sex. Writing "On Narcissism," Freud observed that

> The individual himself regards sexuality as one of his own ends; whereas from another point of view he is an appendage to his germ-plasm, at whose disposal he puts his energies in return for a bonus of pleasure. He is the mortal vehicle of a (possibly) immortal substance—like the inheritor of an entailed property, who is only the temporary holder of an estate which survives him.[3]

People feel attached to their bodies. And well they might. Assum-ing that the Hindus have missed the boat and reincarnation is flat-out wrong, then our bodies are the only "us" we have ever known. More-over, unless truly bionic people come to pass, our bodies are the only "us" that we will ever be. And so, skin-encapsulated egos that we are, most people tend to resist the notion that their bodies are less than all-important. Maybe, in fact, it is "only natural" to rebel against a scien-tific world-view that considers each of us nothing more than the unwitting victims of a multi-billion year old hostile take-over, the longest running version yet of "The Invasion of the Body Snatchers."

In his book, *Darwin's Dangerous Idea*, philosopher Daniel Dennett has examined the widespread resistance to evolution, concluding that many people—even many scientists—harbor a deep desire for "skyhooks." After all, there is a lot of work to be done, in the process of constructing, from mud and blood, matter and splatter, something as elaborate as a human being. That's where the skyhooks come in. Think of imaginary contrivances that somehow attach to the heavens and enable all sorts of heavy intellectual lifting to take place, independent of the real-world constraints imposed by such troublesome and mundane considerations as, for example, friction or gravity. Skyhooks are "miraculous lifters, unsupported and insupportable," but much desired by those who insist that the living world in general, and human beings in particular, could not possibly be fully explained by the operation of simple, brute, mechanical—or organic—processes.

Continuing his construction analogy, Dennett contrasts the yearning for skyhooks with the reality of "cranes," which "can do the lifting work our imaginary skyhooks might do, and they do it in an honest, non-question-begging fashion. . . . [Cranes] have to be designed and built, from everyday parts already on hand, and they have to be located on a firm base of existing ground. . . . Cranes are no less excellent as lifters, and they have the decided advantage of being real."[4]

Billiard balls, cell metabolism and the laws of planetary motion are cranes; extrasensory perception, telekinesis and astral projection are skyhooks. Evolution by natural selection is a crane; special creation is a skyhook. Genes are cranes; souls are skyhooks. Cranes are real; skyhooks aren't. Because they are real, cranes can also be grubby, messy, and sometimes frightening. Because they are ideal, skyhooks tend to be beautiful, perfect, and reassuring. Like the *deus ex machina* of ancient Greek drama, or Superman of twentieth-century comic books, skyhooks can be relied upon—at least in theory—to save us whenever the going gets tough.

Interestingly, most people who believe in an afterlife don't feel diminished by considering that their bodies are in some way inhabited and transcended by something known as the "soul." There is, in fact, an interesting parallel between current understanding of genes and the widespread belief in the human soul as something that exists somehow independent of our physical selves, and that lives forever. But unlike souls, genes have a demonstrable physical existence. In addition, as I hope to show, genes haven't so much taken over our bodies as created them in the first place . . . for their own selfish benefit. (In some cases, there may in fact be hostility between the interests of the body as a whole and its constituent genetic parts; more of this later, especially when altruism is discussed.)

In any event, one cornerstone of revolutionary biology is that evolution is the key to all life, including human beings. But it goes further: the apple of evolution's eye is the gene, not the body. This gene-centeredness helps explain why blood is thicker than water, why people are what they are and do what they do. It is a matter of matter—that is, cranes—and not skyhooks. But it is neither dreary nor humdrum. Like any perfectly good crane, it rests solidly on the ground, but like the best cranes, revolutionary biology reaches to dizzying heights. When the abracadabra is removed, it is no less astounding, and a whole lot more bracing, to look beyond imaginary skyhooks and see what it really is to be human.

In the Beginning . . .

In the beginning was the gene.

Tickled and stirred by lightning, volcanic heat, ultraviolet radiation, and God knows what else, it sloshed around in a rich broth of chemicals that biologists like to call the "organic soup." Actually, it wasn't a gene yet, just a fancy, complicated molecule, one among many. But with billions upon billions floating about, it seems likely that at least one ended up with the ability to make copies of itself. Nothing unnatural in this: after all, many chemicals can be persuaded to replicate themselves in various ways, making, for example, long molecular chains in test tubes.

If you had been watching in those early days, at some point you would have concluded that our friend the Ur-gene was, well, alive. Still, it wasn't much to look at, this ancestral molecule, just a naked copy-catting chemical, scavenging off the land (actually, the broth), reproducing itself as best it could. As this self-replicator went about its business, it doubtless took advantage of the energy present in the great organic gumbo surrounding it, and used that energy (as well as the molecular materials), to keep itself going. In other words, it ate. It also made additional copies of itself. Again, nothing weird, unnatural or miraculous about this: the likelihood is that many molecules—almost certainly, the great majority—did *not* replicate themselves. Looking back with 20/20 hindsight through the long corridors of evolutionary time, archaic couch-potatoes which didn't replicate (or which did so inefficiently) were among the losers, so much so that they are not even visible to us now. At the time, they probably would not have seemed like losers at all; it is just that any "living" but non-replicating (or poorly replicating) molecules would eventually have been degraded by the slings and arrows of outrageous fortune: over time, they would have been broken down by other chemicals, or just broken up by being knocked about.

The winners were those that not only kept themselves going, but that also left descendants. Mere survival was not enough, just as it isn't enough today. Evolutionary laurels only go to those who succeed in projecting copies of themselves into the future, even if this is done at substantial cost: in pain, time, energy, even shortened life span. Take, for example, the average man. He lives on balance less long than the average woman, for reasons that are only now being unraveled. One possibility is that the male sex hormone, testosterone, makes the resting metabolic rate of men about 5 percent higher than that of women; as a result of running their biochemical engines at a higher speed, men may wear themselves out more quickly. Another possibility is that since maleness means sperm-making, this in turn leads to riskier, more competitive activities. Thus, acting like a male means courting a greater chance of mortality.[5]

Whatever the precise cause, the connection between maleness and higher mortality is undeniable: men who are castrated when young live longer, approaching women in their average longevity. And yet, evolution has not responded by producing men without testicles, nor would we ever expect that to happen! Natural born castrati would be winners in a sense, in that they would almost certainly live longer, but because they couldn't reproduce, evolutionarily they would be losers.

Back in the primordial pea soup, our molecular ancestors found themselves competing with others, trying (although almost certainly not consciously) to be winners rather than losers. In short, to reproduce. But although reproducing was the bottom line, doing so successfully almost certainly required being successful in a number of other, peripheral activities, such as getting a meal for themselves, while also avoiding becoming a meal for something (someone?) else. One way to compete successfully might have been to enclose one's self in armor of sorts, creating a "skin" that we would now identify as a cell membrane. Those who did would have survived better, and also reproduced their successful tactics. Not surprisingly, therefore, their descendants were also likely to build themselves a protective membrane.

These primitive cells, constructed by the earliest living molecules—lets call them genes—for their own perpetuation, were living things in their own right, but at the same time, they were a bit like robots, outfitted with a growing array of prosthetic devices for obtaining food, warding off other cells, regulating their internal chemical environments, and so forth. All for the good of their creators, the genes. Bear in mind that by this point, genes weren't only making copies of themselves; instead, like the CEOs or production managers of a large corporation, they were directing the activities of the increasingly complex structures—the cells—they had created.

A long and exciting journey was underway, from naked, copy-catting molecules inhabiting their frothy organic soup to sophisticated modern genes pulling the intricate levers of control in today's multicellular creatures. On route, all sorts of evolutionary experiments were tried. One of the more momentous was the switch from one-celled organism to larger, multicellular creatures. This could have happened in many different ways, such as when a single cell, after dividing, remained stuck together. Such accidents, especially after being repeated many times, would have offered an entrancing new possibility: specialization, in which some cells became experts in, say, obtaining food while others specialized in digesting it, or in providing locomotion for the whole assemblage. Everyone gained in this early division of labor, a specialization that probably did not elicit a great deal of conflict, since if they started off as genetically identical "daughter" cells, the different cells would all have been composed of identical genes. As a result, whenever the "body" profited, all the underlying genes benefited equally.

The creation of bodies would have been just one of many early developments. Others include various experiments with reproduction, notably the discovery of sex, whereby genes from one organism combined with genes from a partner and then rearranged themselves to produce offspring who were truly *new*, that is, who represented gene combinations that were different from those of their parents.

Note, however, that in this process the genes themselves did not become new; they just found themselves paired with new partners. Actual changes in the genes only happened when a mutation—that is, a copying error—took place. Mistakes are inevitable; they happen. Mostly, these inevitable copying errors were hurtful, even disastrous. Bear in mind that even the earliest organisms were pretty complicated by comparison with the organic soup in which they swam (not to mention the nonliving environment which, by contrast, was so simple as to be downright boring). Since the initial copying errors were almost certainly random—as most mistakes are—they generally made things worse rather than better.

Analogously, imagine what would happen if you removed a few ounces of metal from an automobile, or computer, and replaced it with a randomly chosen handful. Every now and then, however, one of these random errors would have helped more than it hurt; such benevolent "errors" would have made their carriers *more* successful, rather than less, and would have therefore become incorporated into the collection of successful genes. (I made an equivalent error some time ago, while writing an article and intending to type "man-made world." Instead, the result was "mad-made world," which turned out, quite by chance, to be an improvement. In this case, the chances of such a random event actually making things better were 1 in 26: max-made world wouldn't have "survived," nor would may-made world, etc. In the case of analogous genetic errors, the chances appear to be between 1 in 100,000 and 1 in one million: Rare, but given enough time and enough genes, reliably frequent.)

Such mutations would have added new genetic materials to the available supply, simply because they were able to out-compete their rivals. Eventually, living things came to be composed of genes that—in evolutionary terminology—were "adapted" or "fit," which simply means that they were good at creating bodies that, in turn, did a good job of meeting the problems of life.

The early stage of life was undoubtedly a time of ferment and experimentation, just as it is today. Some genes were more fit when they produced bodies that were small and furtive; others, when their bodies became large and imposing. Some genes profited by creating bodies that were especially resistant to drought, or well camouflaged to match their backgrounds, or inclined to cooperate with particular bodies while beating up others.

As the evolutionary process gained momentum and genes became ensconced within "their" bodies, an objective observer wouldn't have seen genes at all, just bodies going about their business, living their lives, cooperating with some, competing with others, some winning, some losing, some marking time. Behind the scenes, however, things had not changed, *and still have not changed.* Inside the bodies and beneath the surface is a never-ending sequence of yanking and tugging at the cellular and organismic levers of power, all of it going on for the same original, unchanged reason: genes working to make copies of themselves, by hook or by crook.

They Endured

Early in his novel, *The Sound and the Fury*, William Faulkner pronounces this judgment upon his seemingly doomed, yet stubbornly persistent characters: "They endured." The same can be said of genes.

The Compson family of Mississippi endured through lust and violence, madness and incest, thievery and the heavy hand of history; ditto for genes. Genes also endured through drought and flood, predators and parasites, friends and foes, good times and bad. And unlike Faulkner's fictional inhabitants of Yoknapatawpha County, genes are real.

But of course, not all genes endured equally; that is what is meant by natural selection. As already stated, one of the simplest ways (although as we shall see, not the only way), for some genes to outendure others to create bodies that are more successful, either because they are better constructed, better functioning, or because they behave in ways that are more effective. Most of the time, therefore, the ultimate interests of genes coincide with those of the bodies they create. After all, that's why genes create bodies in the first place: to look after their interests. It is also why biologists have long been able to pursue their work, to ask interesting questions and get interesting answers, even before the evolutionary revolution, which provided convincing evidence that genes count for more than bodies. (As we shall see in the next chapter, it is when the interests of genes and bodies do *not* coincide that things get even more interesting, and when the surprising importance of genes is highlighted.)

Not that the genes—or the bodies they create—consciously tell themselves that gene-endurance is their ultimate goal. There is no saying what, if anything, they say to themselves or their cellular slave-mechanisms. The important thing is that those genes which (who?) directed their bodies to act effectively—to avoid predators, obtain food, stay out of dangerous places, etc.—eventually inherited the earth (along with the sea and the air), because their bodies prospered. And this, in turn, enabled the genes to do the same.

But again, don't forget that "prospering" or enduring for genes isn't necessarily the same as "prospering" or enduring for bodies. For genes to do well—that is, to be selected—they must project themselves down the river of time, and the best way to achieve such a trip is to violate the old adage and change "horses" (that is, bodies) in mid-stream. In this case, however, it is less a matter of abandoning their body than of creating a new one, and outfitting it with copies of themselves. That is, to have children. As vast, almost unimaginable time passed, some of these early genes and the cells and bodies they built were more successful than others, not because they had particular insight, but just because they created effective bodies, which—as a crucial part of their success—were able to spin off new bodies every now and then. Within these new bodies were copies of themselves. These gene-carrying bodies are known as offspring. And they are an important key to the endurance of genes.

This has now become the bottom line when it comes to biological success: reproduction. Those individuals who are better at it—that is, better at propelling their genes into future generations—are more fit, and their descendants (sometimes meek, sometimes mighty, always enduring) have inherited the earth.

Survival of the fittest? Yes indeed. The next question, then, echoes Richard Alexander: The fittest what? In the short run, the fittest individuals, if only because genes cannot even get started on their evolutionary way if their bodies perish too soon, carrying genes down to defeat along with them. In the long run, however, it is genes and their success that matters. And the bodies, everything that they are and do, every way that they strive and hope, exult and despair; from the vantage of evolution, these are nothing more than means to an end.

What end? Endurance. It is troublesome but true that in another sense, mere continuance (even increase) isn't really an "end" at all. It is really just a matter of staying in the game.

As the early life-forms kept on keeping on, in some cases becoming more abundant, they would have devoured their surrounding nutrients, probably starting with those other complex, not-quite-so-enduring molecules which were not up to the task of making successful copies of themselves, not to mention constructing bodies to do their bidding . . . and thus, were not sufficiently "alive." But then, trouble almost certainly arose: not enough food. The soup grew thinner, less like gumbo and more like consommé.

Faced with what in all probability was a major crisis, perhaps *the* greatest crisis in the history of life, some of these life-forms eventually hit upon a creative solution: make their own food. Instead of relying on the increasingly slim pickings around them, those genes that enabled their bodies to capture energy from sunlight and use it to synthesize their own food would have been at a great advantage. Enter, the first plants.

More like primitive algae than like rose bushes, they would have provided a food source for the other life-forms that never discovered the benefits of chlorophyll. These primitive grazers on the proto-plants were the archeo-animals, the world's first genuine herbivores, and they, in turn, became food for other life forms who preyed upon them; namely the first true carnivores. As Kurt Vonnegut might have said, "And so it goes."

As it went, with genes mutating and enduring (and sometimes staying the same and perishing), living things became more diverse, branching out to occupy all sorts of niches, filling the oceans, lakes, and rivers, eventually colonizing the land and even—to a lesser extent— the air. Some evolved hair and a novel way of nourishing their offspring (via milk), thereby becoming mammals. Some of these mammals (the primates) climbed into the trees, whereupon others (early human beings) climbed back down again, for reasons that are still obscure. In any event, they found themselves endowed as a result of their arboreal sojourn with good binocular vision and grasping hands with opposable thumbs. Somewhere along the line, they also evolved large brains, perhaps because with their arms freed to use tools, natural selection favored those who were especially good at doing so. Or perhaps selection favored the distinctive human brain because for weak-bodied primates to survive—never mind prosper—on the prehistoric savannas, they had to be unusually smart. After all, our ancestors were neither as strong nor as fast nor as well defended as most of those other animals with whom they competed. So, perhaps selection smiled upon those whose brains were able to tip the evolutionary scales in their favor. In a perverse way, therefore, the physical *inadequacy* of human beings may have been responsible for unusual intelligence. Or maybe early aggressiveness conferred an advantage on those human ancestors who were clever enough to put up a good defense (or a good offense), against others of their species. Perhaps big brains evolved along with complex social relationships, enabling individuals to recognize other individuals and to vary their responses depending on what served their needs.

But we are not concerned here with detailing the early evolution of life, or even the precise ancestry of human beings. We simply want to emphasize that *Homo sapiens* is a perfectly good mammal, a biologically evolved creature whose pedigree extends every bit as far back into the earliest organic soup as does the ancestry of a giant tortoise, or a slime mold. In the great scheme of things, none of us—that is, no individual body—will endure. Our genes, on the other hand, already have. And they have the potential to keep doing so.

The Meaning of Life?

In our basic structure and physiology, there is absolutely nothing that is organically unique about human beings. We are made of cells, with all the various biochemical and electrical mechanisms and processes that operate within other living things. We are made of genes (or rather, *by* our genes) and even a detailed examination by the Human Genome Project has not found a single gene mechanism that is unique to *Homo sapiens*. (Of course, there are numerous genes found only in human beings, just as there are many genes found only in oranges, or in orangutans; they are what makes oranges oranges, and orangutans, orangutans. The point is that the process whereby genes influence their bodies is the same in all living things, *Homo sapiens* included. We cannot fall out of this world.)

To some extent, every animal and plant species is unique. That's how we recognize it as different from another. But there is nothing unique in being unique! Beyond what might be called the "expected uniqueness" of every distinct life-form, there is nothing in the physio-chemical-electrical-structural make-up of human beings that breaks the rules, or constitutes a radical discontinuity from the life around us.

If there is a single take-home lesson to be derived from the work of Darwin through the outset of the twenty-first century it is just this: continuity. People are natural, organic, biological, evolved critters, just like every other natural, organic, biological, evolved critter on this planet. And as we go about living our lives, we act out the same basic goals, each a modern-day descendant of those early life-forms that distinguished themselves via the endurance of their genes.

The English romantic poet John Keats was no biologist, but in his "Letters," nearly 200 years ago, he seemed to understand. "I go amongst the fields and catch a glimpse of a stoat or a fieldmouse peeping out of the withered grass," Keats wrote. 'The creature hath a purpose and its eyes are bright with it. I go amongst the buildings of a city and I see a man hurrying along—to what? The creature hath a purpose and its eyes are bright with it."

Fifty years before Darwin, Keats realized that animals have a purpose of some sort. And furthermore, he sensed that human beings, too, have a purpose. Although Keats didn't state unequivocally that humans and animals share the same purpose, revolutionary biologists are now prepared to do just that. Furthermore, we can even identify that purpose: achieving the greatest possible success of their genes. (Whether it brightens their eyes is another question.)

Don't misunderstand: people can go about their lives in pursuit of their biological "purpose" even if they don't know what that purpose is, or don't realize they are doing so. Just like the stoats or field-mice that brightened the eye of John Keats. Or like the first gene-molecules of the early soup. If we eat when hungry, sleep when tired, engage in sex when the right opportunity presents itself, run from enemies (or overpower them, or hide from them, etc.), we are behaving purposefully, with that "purpose" going beyond the immediate and obvious motivation of filling our bellies, slaking our desires, and so on. Behind the superficial facade of satisfying one's needs or responding to one's fears, lies the deeper purpose, of satisfying the ever-present, bottom-line requirement of evolutionary success. It is this requirement that has produced such short-term motivators as hunger, fear, lust, and so forth.

In this sense, life really does have a meaning, a purpose. (Recall the Faulkner Principle: "They endured.") Genetic endurance is a meaning and purpose that is shared with the lowliest of creatures. No planning, no foresight, no skyhooks are needed. Not even a brain.

Take trees: about as brainless as you can get, but nonetheless purposeful in the evolutionary sense. In the United States, for example, trees flower in the spring. Pretty clever of them, since that is when insects are buzzing around to provide pollination, and also when the weather is getting warmer so that their seeds, which develop from the flowers, will have a good chance of success. Henry James referred to life's "blooming, buzzing confusion," a felicitous circumstance to which plants contribute by their blooming, and insects, by their buzzing. But this happy concordance is neither coincidence nor the result of conscious design by either plant or insect. Plants and pollinating insects bloom and buzz at precisely the season when doing so will maximally reward the genes that are responsible for all that glorious blooming and buzzing.

Natural selection is the process by which certain genes are chosen over their alternatives. Imagine a set of genes that caused "their" bodies (in this case, trees) to flower during, say, December. Imagine, also, alternative sets of genes that caused "their" bodies to flower during other months. The December genes would have this effect: trees that housed them and that followed their ill-fated genetic promptings would have come to an inglorious and unfruitful end, in the snow, rain, and darkness of winter. Moreover, as the bodies go, so the genes would go, too, dragged along by their own unfortunate effects. By contrast, genes that induced flowering in April (or perhaps May, or June, depending on the species and its circumstances) would hit the jackpot. They would leave more descendants, so that eventually, people might find themselves marveling at the wisdom of trees: altogether devoid of brains, yet flowering when they should, as though they knew what they were doing.

Trees are not alone in being ignorant of biology and yet behaving with sophisticated evolutionary wisdom. There are places on earth whose human inhabitants supposedly did not understand, until recently, the connection between sexual intercourse and reproduction. And yet, they managed to reproduce. And there are people, today, who refuse to understand that human beings are inclined to behave in ways that benefit their fitness (that is, the success of their genes) even if they don't consciously know anything about fitness. Even these skeptics, however, aren't troubled by the fact that trees flower at the right time of year without knowing anything about the relations between spring, warmth, pollination, and the probability that their seeds will germinate.

That's all right for trees and bees, but what about human beings? Haven't we risen above such mindless constraints? For an answer, take a moment and put your finger in the upper right corner of your mouth; feel around on the gum, just above the teeth. Notice that curious little bump? Now move your finger to the upper left corner and likewise the lower right and lower left. Each of those little protuberances on the four corners of your jaw is known as a "canine eminence," and it is a small vestige of our shared evolutionary past. The significance of the canine eminence becomes clear upon examining the dentition of a baboon, gorilla, or chimpanzee (from a safe distance!): you can hardly miss the large, fierce canine teeth, which require substantial anchoring.

Each of us carries many such bodily vestiges of our evolutionary history. Also behavioral vestiges, many of which are not merely irrelevant leftovers, like the canine eminence, but crucial to our current lives. Take another moment, therefore, and listen to the singing of birds in the spring. Then, listen to your own internal singing: at spring, at the satisfactions of love, of parenthood, of a good meal, or a good sleep . . . even a good bowel movement. Here is naturalist Donald C. Peattie, responding to the springtime chorus of frogs:

> It speaks of the return of life, of animal life, to the earth. It tells of all that is most unutterable in evolution—the terrible continuity and fluidity of protoplasm, the irrepressible forces of reproduction—not mythical human love, but the cold batrachian jelly by which we humans are linked to things that creep and writhe and are blind yet breed and have being.[6]

Hardly anyone talks about "protoplasm" any more; its about as useful as identifying the contents of a computer as "cyberplasm." And "batrachian" has a distinctly quaint sound as well. But there is nothing withered or outdated about the point Mr. Peattie is trying to make. It is, instead, a matter of blood and guts, sperm and eggs, the fervent stuff of seething, surging life itself.

"We be of one blood, thee and I." That was the secret pass-code learned by the feral child, Mowgli, in Rudyard Kipling's *The Jungle Book*. By it, Mowgli gained the allegiance of all creatures of the tropical Indian jungle. Mowgli and his friends are fiction, but in a literal sense, we really *are* of one blood: frogs, fishes, and fowl. Birds of a feather, all of us, all sharing an evolutionary history. More to the point, we all share genes, notably those that underpin the gears and levers controlling the basic operations of aliveness: how energy is metabolized, how and when chromosomes divide, how DNA oversees the manufacture of proteins, and so forth. The same genes, (or, more accurately, identical copies of the same genes) have created our fellow human beings no less than fellow frog, batrachian jelly and all. And insofar as these genes help themselves to achieve evolutionary success by inducing their bodies to behave in particular ways, most bodies, most of the time, whether frog or falcon, porcupine or person, are likely to go along.

This is not to say that human beings are "nothing but" animals; clearly, we are, as the slang expression puts it, "something else." People are animals *plus*. Some believe this "plus" is a soul; most biologists believe that our species' specialness resides in humanity's rather over-sized brain, and that language, imagination, culture, symbolism, self-consciousness—all result from big-braininess. But however you slice it, no Martian biologist worth his or her salt, upon studying the life-forms of planet Earth, would grant human beings status as non-biological, or in any fundamental way extra-biological. In the deepest, most profound sense, everyone is like everything else. Of one blood: you and I, and everyone.

In his novel, *Timequake*, Kurt Vonnegut's alter ego, science fiction writer Kilgore Trout, observes that "We are here on Earth just to fart around. Don't let anybody tell you any different." Vonnegut/Trout are quite correct that "we are here on Earth" for no cosmic reason, but neither is it our mission to "fart around." Rather, it is to promote those genes that made us.

The gene's-eye view of evolution was described most dramatically by the English evolutionary publicist Richard Dawkins, in his much-heralded book, *The Selfish Gene*.[7] In it, Dawkins described genes as "replicators," emphasizing their most important trait: the ability to copy themselves. (Second in importance would be their ability to construct bodies and equip those bodies with instructions as to how to grow and develop, as well as how to behave under different circumstances.) Dawkins also emphasized that what are known as bodies, or organisms—that is, individual plants or animals—are really "survival machines" manufactured by genes for their own selfish benefit:

> Replicators began not merely to exist, but to construct for themselves containers, vehicles for their continued existence. The replicators that survived were the ones that built survival machines for them to live in. The first survival machines probably consisted of nothing more than a protective coat. But making a living got steadily harder as new rivals arose with better and more effective survival machines. Survival machines got bigger and more elaborate, and the process was cumulative and progressive.

Was there to be any end to the gradual improvement in the techniques and artifices used by the replicators to ensure their own continuation in the world? There would be plenty of time for improvement. What weird engines of self-preservation would the millennia bring forth? Four thousand million years on, what was to be the fate of the ancient replicators? They did not die out, for they are past masters of the survival arts.[8] But do not look for them floating loose in the sea; they gave up that cavalier freedom long ago. Now they swarm in huge colonies, safe inside gigantic lumbering robots, sealed off from the outside world, communicating with it by tortuous indirect routes, manipulating it by remote control. They are in you and in me; they created us, body and mind; and their preservation is the ultimate rationale for our existence. They have come a long way, those replicators. Now they go by the name of genes, and we are their survival machines.

Surprisingly, after this effective, if overheated prose, the author backed off when it came to human beings. *Homo sapiens* disappeared almost entirely from the rest of his otherwise excellent book. Maybe Professor Dawkins experienced a sudden and uncharacteristic burst of intellectual modesty when it came to his own species. I am not so shy.

Genes "for" This and That

Practitioners of gene-centered revolutionary biology refer unblushingly, without quotation marks, to genes "for" all sorts of things: altruism, selfishness, even tendencies to argue with one's parents (and offspring). But what does it mean to talk about a gene "for" anything, behavior in particular?

Isn't it awfully skyhookish?

Not at all. Genes aren't mystical or imaginary, the modern incarnation of phlogisten, caloric, or the interplanetary ether. Neither are they "social constructs" or the results of ideologic spectacles of one sort or another. Genes are cranes, real physical stuff, down-to-earth concrete organic molecules, conglomerations of the chemical known as DNA and consisting of various patterns of four different building-blocks, specifically molecules known as adenine, thymine, cytocine, and guanine, each of them composed of known arrangements of atoms.

Moreover, although some mystery continues to surround many of the fine points of how genes actually go about their business, the basic outlines of gene action are clearly understood, even as additional details are rapidly being revealed. In short, genes are instruction manuals, carrying information that informs each cell to produce its allotment of proteins, which in turn is crucially important as the basic infrastructure of life, and also as the enzymes that control the various, complex reactions that make living things tick.

Genes undeniably carry the information necessary to produce, for example, your heart, your muscles, and your skeleton. In exactly the same way, genes carry the information necessary to produce your brain. There is, in short, no fundamental difference between genes encoding the structure of the heart, and genes encoding the structure of the brain. Finally, to complete the connection: Just as we all know that the pumping of blood results from the beating of the human heart, there can be no doubt that behavior—as well as emotion, thought, self-awareness, etc.—results from the working of the human brain. (Just as the behavior of other living things results from the working of *their* brains, if they have any.)

There is, admittedly, something incongruous about equating the wonder and glory of human thought, emotion, and consciousness with the squirting of blood through the aorta. And indeed, the so-called "mind-body problem" is one of the oldest and thorniest in both science and philosophy. How can anyone bridge the gap—indeed, the chasm—between neurons and symphonies, between electro-chemical potentials and love, pity, despair, or exaltation? It may be one of those things that is difficult—maybe impossible—in theory, yet easy in practice! Each of us does so, every time we think, hear, feel, imagine, and hope.

The fact that genes influence behavior (via the brain) is in the same conceptual ballpark as electro-chemical events influencing consciousness, or other genes influencing height or eye-color (via skeletal structure, or pigmentation). To be sure, consciousness is not a physical object or a slimy secretion, and yet, human thought is produced by the human brain, just as urine is produced by the kidneys.

Does this mean that there is a gene "for" every behavior? For tying your shoelaces, enjoying music, loving children, defending your country? Almost certainly not. After all, it now appears that there are only about 100,000 genes in each human being, and it would be easy to identify more than 100,000 different behaviors alone, not to mention all the other things (structure and function of the heart, kidneys, etc.) about which genes have their say. The truth is both simpler and subtler than "one gene = one behavior."

Behavior is not contained within genes, ready to pop out fully formed, like the goddess Athena emerging from the forehead of Zeus. Rather, behavior develops over time, under the joint prodding of genetics *and* experience. It is therefore misleading and just plain untrue to maintain that genes, by themselves, "control" behavior, any more than they single-handedly "control" the making of livers or kidneys. And just as there is no single, simple gene "for" livers or kidneys, there is no single, simple gene "for" behavior. But at the same time, *changes* in certain genes can cause *changes* in livers or kidneys, just as changes in other genes can cause *changes* in behavior.

Almost daily, genes are discovered that predispose people to such diseases as breast cancer, Alzheimer's, and so forth. In such cases, the suspect genes are not solely responsible for the outcome. Instead, individuals possessing them are somewhat more likely to develop the disease than people with other genes. A purported gene "for" a disease like Alzheimer's or "for" a behavior like altruism doesn't "cause" Alzheimer's disease or altruism all by itself. It acts in conjunction with other things, which, in the case of diseases, are known as "risk factors," and in the case of behaviors, as the "environment."

In the field of animal behavior, there is simply no question that genes influence behavior, just as they influence other traits. The study of "behavior genetics" is as well established as other scientific disciplines, such as physiology, anatomy, or biochemistry. Some examples:

There are two species of African lovebirds known as *Agapornis*. One species carries nesting material in its bill; the other, by stuffing bits and pieces into its rump feathers, and then flying back to its nest with the stuff flapping behind. Biologist William Dilger of Cornell University got the two species to interbreed, whereupon he found that the hybrids were genuinely confused: given strips of paper, they would begin to pick them up in their bill, pause as if undecided, then stick some of them in their rump feathers, then pull these out and hold them for a time in their bill once again, and so on. Clearly, genetic tendencies to do two contradictory things were vying within the mixed-up brains of these genetically mixed animals.

By choosing laboratory mice who are especially fond of ethyl alcohol, behavior geneticists have been able to produce, via artificial selection, strains of mice that prefer alcoholic drinks to plain water, and higher proof to lower (or vice versa). In 1996, a team of developmental geneticists created so-called "knock-out mice," individuals who were lacking a specific gene, in this case one that codes for the release of the brain chemical serotonin. As a result of this single genetic change, such individuals preferred 20 percent alcohol over tap water, and they consumed, on average, twice as much alcohol as normal mice. Lacking this single gene, otherwise normal mice were transformed into alcoholics.[9]

Long before modern laboratory studies, the evidence for behavior genetics was overwhelming, even when it wasn't identified as such. Beginning thousands of years ago, long before anyone knew DNA from Darwin, animal breeders began "creating" strains of animals that differed from the wild type, and from each other. Via artificial selection (allowing individuals with certain traits to reproduce disproportionately, and keeping this up for many generations), these early practitioners of animal husbandry gave us cattle that are more docile than their wild ancestors and that produce much more milk, hens that lay lots of eggs and roosters that fight like crazy, pigs that are inclined to hang around and grow fat rather than run around and get skinny, horses that are exceptionally high-strung (such as Arabians) or easygoing (known today as warmbloods), and so forth. From the ancestral wolf, there are now hundreds of different breeds of dogs, which differ from each other behaviorally no less than in appearance: a golden retriever is significantly less liable to bite than is a rotweiller and also far more likely to retrieve. Some dogs—for example, Shelties—love to herd, while others, such as terriers, are passionately fond of going into tunnels in search of rats. Some breeds don't even bark. Others won't stop.

Dogs can be trained to have a sense of guilt (or perhaps, to anticipate punishment). But not all breeds are equally susceptible. In a group of experiments in which dogs of two different breeds were punished for touching some meat and were then left alone with it, it was found that Shetland sheepdogs refrain from touching the forbidden fruit, whereas Basenjis eat the meat as soon as the trainer leaves the room.[10]

In all these cases, the reason is genes; or, to be more accurate, the interaction of genes with environment. Another way to look at it: the reason members of one breed differ, on average, from members of another is that certain genes have been substituted for others.

The same thing applies to individuals. Among the tiny roundworms known to science as *C. elegans*, there are two strains that vary in their tendencies for sociality, especially while feeding: Individuals of one strain are social eaters, the others prefer to be solitary. The difference is determined by alternative forms of the same gene. At the same point on a particular DNA sequence, social worms have T-T-T (producing the amino acid phenylalanine), whereas solitary worms have G-T-T at the corresponding position (which codes for the amino acid valine).[11]

Scientists at the National Human Genome Research Institute, led by clinical geneticist Anthony Wynshaw-Boris, have discovered a single gene that exerts a profound effect on the social behavior of laboratory mice.[12] It was discovered by creating individuals lacking the gene in question (known as "disheveled-1"). These mice, with the gene in question "knocked out," engage in fewer social interactions than normal. They also spend less time huddling together, make inadequate beds out of nesting material, and fail an all-important mouse duty: trimming one another's whiskers. (Typically, in the strain of laboratory mice being studied, dominant individuals carefully barber the whiskers and facial hairs of their social inferiors.) Also interesting is the fact that knockout mice lacking the disheveled-1 gene seem unable to screen out extraneous noise, and therefore have difficulty focusing on a single stimulus. Their disability may in turn be related to several human psychiatric disorders, including schizophrenia, Tourette's syndrome, autism, and/or Attention Deficit Disorder.

Literally thousands of individual genes have been identified in insects, fish, birds, and mammals, each of them influencing one behavioral trait or another. In fact, it is virtually unheard-of for a behavior geneticist to attempt to select for any behavior in any species of animal, and to fail. (Once again, this does not necessarily mean that there is a gene "for" every behavior; rather, it suggests that by changing the genetic background, it is possible to alter any behavior.)

What about a gene "for" enjoying roller-coasters, say, or for mountain climbing? Sounds absurd, and in a sense, it is. But consider this: In January, 1996, two different research teams—one from Ben-Gurion University in Israel, led by Richard Ebstein, and the other from the National Institutes of Health in Bethesda, Maryland, led by Dean Hamer—reported that they had discovered a gene for "novelty seeking" in people. (It is especially impressive that the finding was replicated in two very different populations, Ashkenazi and Sephardic Israelis of both sexes, as well as Caucasian American men.)[13] The newly described gene provides instructions to the so-called D-4 dopamine receptor in the human brain, which, in turn, influences how its possessors react to new and—depending on one's perspective (plus genetic make-up)—"exciting" stimuli. If you were to examine the DNA of roller-coaster aficionados, you would likely find a higher-than-average proportion of this "novelty-" or "thrill-seeking gene," as opposed to the stay-at-home, couch potatoes among us, who, one may assume, are disproportionately drawn from those possessing the alternative, more sedentary form of this particular pattern of DNA.

The case of novelty-seeking was the first example of an identified human gene concerned with a normal behavioral predisposition, rather than a structural defect, or disease susceptibility. Almost certainly, it will not be the last. It also serves as a useful metaphor for how genes influence behavior: by working on inclinations and proclivities, rather than commanding a robotic, lock-step insistence on doing a particular thing. Undoubtedly, there are roller-coaster fans and avid mountain climbers who lack this particular genetic variant, just as there are scaredy-cats and stay-at-homes who possess it. But across the human population as a whole, it is a good bet that the gene and "its" behaviors coincide.[14]

Not long after the novelty-seeking gene, the genetic underpinning of another "normal" behavior was reported, this one dealing with equanimity versus neurotic anxiety, and known as the "5-HTT promoter." It operates by altering the body's use of a different neurochemical, serotonin, long known to be involved in depression, anxiety, and lack of confidence.[15] Once again, of course, people may well be "worry-warts" even if they lack the identified genetic variant, whereas some perfectly confident individuals may have it. But a genuine correlation appears to exist, one that is likely to be causative as well. Not only that, but it is likely to have ramifications in other aspects of daily life.

On average, people with a greater-than-usual tendency to worry may have a correspondingly greater-than-usual tendency to drive particularly crash-resistant cars, to buy extra life insurance, or to invest in government bonds rather than the stock market. And on the whole, these people would likely have a greater-than-usual probability of carrying the 5-HTT promoter gene. In short, the 5-HTT promoter gene is "for" buying Volvos![16]

Take an even more mysterious and vexing case, homosexuality. Here we see the subtle interaction of genes and experience. Among the Sambia, a Melanesian people of the South Pacific, there is one and only one form of sexual behavior condoned for boys: the younger ones are encouraged to suck the penises of older boys, thereby, it is thought, gaining masculinity from their semen. If experience counts for everything, then these young people should grow up homosexual. And yet, most young Sambian men become fully heterosexual upon marriage; homosexuality is no more frequent than among other people. One likely lesson from this admittedly unusual case is that social learning and experience neither erase heterosexuality nor instill homoerotic tendencies. (At least, not by themselves.) What does, then?

One possibility is that each individual's genetic make-up predisposes him or her to choose sexual partners of one sex or another. This is not the same as saying that homosexuality—or heterosexuality, for that matter—is "in" the genes; rather, genetic predisposition, plus a complex array of experiences, interact to produce each person's sexual orientation.

Such genetic predispositions are rapidly being uncovered. In a study of 161 homosexual men with brothers, both brothers were found to be homosexual 22 percent of the time, whereas in the case of adoptive "brothers"—who experienced the same home environment but were genetically unrelated—both men were homosexual only 11 percent of the time. This in itself provides strong presumptive evidence for a genetic role. Even more convincing was the finding that when identical twins were examined, both were homosexual 52 percent of the time. Similar findings have been reported for lesbianism.[17]

For a final example of genetic impact, a study of identical twins aged eighty and older compared them with comparable-age "fraternal" twins (who are no more closely related than any pair of siblings). The researchers found that members of the identical group were substantially more similar to each other—not just in physical appearance but also for traits such as general cognitive ability, verbal ability, spatial ability, speed of information processing and memory—than were pairs from the fraternal group.

Results of this sort, although important, are not surprising: They have been well known for decades. New, however, was the discovery that the influence of genes on cognitive ability is not only profound, but remarkably steady throughout life. Thus, it had been thought that genetic influence was especially strong early in development, but was eventually overwhelmed by the vagaries of personal experience, a kind of idiosyncratic wedge that widens the gap between individuals as time goes on. Not so: eighty-year-old identical twins eerily persisted in resembling each other with regard to their intellectual performance, just as eight-year-old twins do.[18]

This is not to say that genes "control" human behavior, that there are any modern-day Calibans, "on whose nature," as Prospero says in Shakespeare's *The Tempest*, "nurture can never stick." Nurture influences everyone's nature, just as nature—that is, genetic makeup—influences everyone's response to experience.

My youngest daughter is a violinist. When she plays a Bach sonata, is it Nellie who makes the music? Or her violin? Neither, alone. To be sure, the violin matters: the sound produced is very different if Nellie plays on a toy, or a Stradivarius. So does the violinist: Nellie's own violin would sound different indeed if played by Itzhak Perlman. In the end, it is both musician *and* instrument who make the music. More precisely, it is the *interaction* of the two.

Take another behavior that is obviously influenced by one's surroundings, such as reading. There is no doubt that proficiency at reading is affected by direct experience: some people, after all, are perfectly capable of reading but can't read, because they never *learned*. Reading is also affected by practice, motivation, opportunity, vocabulary, even such trivial factors as lighting, quiet, and a comfortable place to sit down. Nonetheless, there is also genetic variation with regard to reading ability. Some people are dyslexic; they have difficulty—a difficulty that is genetically influenced—making sense of symbols organized in linear sequence. This does not mean that there are genes for reading in the same sense that there are, for example, genes for smooth vs. wrinkled peas. But undoubtedly there are genes that influence visual acuity, hand-eye coordination, retinal structure, color-blindness, connections in the cortex . . . anything that, once inputted, affects one's ability to derive meaning from squiggles of printer's ink. So, there are genes "for" reading—in that reading is influenced by their presence or absence—just as there are genes "for" buying Volvos or enjoying roller-coasters even though there are no genes that directly control any of these behaviors.

From here on, we shall therefore remove the quotation marks when considering genes "for" behavior, even complicated behavior.

The Genetic Sweet-Tooth

The second chapter of Edward O. Wilson's monumental book, *Sociobiology*, began with this enigmatic observation: "Genes, like Leibnitz' monads,[19] have no windows; the higher qualities of life are emergent." Since they lack windows, the only way to see into the inner workings of genes is by their outer manifestations; namely, the bodies they create and manipulate. Genes are the prop designers, stage managers, and directors, but bodies are the actors, the ones in the spotlight. When it comes to behavior, the whole is greater than the sum of its parts, even though to understand the whole we need to understand the contribution of the parts.

And so, despite the importance of focusing on genes as the basic unit of evolution, throughout most of this book we will nonetheless talk about the behavior of individuals, meaning whole, intact human beings. After all, even though our genes ultimately call the evolutionary tune, our bodies R us.

These bodies do all sorts of things, things that on balance benefit the genes that created them, and which, not surprisingly, are perceived by those bodies as being pleasant. People, after all, are not indifferent to their actions. They generally prefer doing some things to others. And although there is no easy accounting for the details of taste (some like milk chocolate, some bittersweet), evolution helps to account for most human preferences, whether in food or other aspects of life.

Let's continue the metaphor: Certain tastes are widespread. Take sugar, for example. Why do most people like it? In other words, why is sugar sweet? After all, here is a widespread taste preference, for which there is presumably a genetic underpinning and some adaptive (that is, evolutionary) significance, even though it often leads to disadvantages such as obesity and tooth decay. Presumably the perception of "sweetness" has something to do with the sugar molecule, and yet, dogs and cats, for example, do not especially like sugar. For them, sugar is not sweet. "Sweetness." then, does not simply reside in the chemical known as sucrose. Anteaters, if asked, would almost certainly point out that ants are sweet! If we were intelligent anteaters, perhaps we would be asking whether sweetness resides in formic acid or in some other constituent of our favorite food, whereupon some anthropologically inclined anteater might astound its colleagues by pointing out that human beings, peculiar creatures that they are, consider *sugar* to be sweeter than ants!

Is there, then, *no* accounting for tastes? Quite the contrary. The likelihood is that people find sugar sweet because they are primates who evolved as fruit-eaters, and for their ancestors, sugar indicated ripeness and thus, nutrition, just as anteaters find ants sweet because they have evolved to make good use of ant-stuff (as giant pandas have evolved to make good use of bamboo-stuff, etc.).

People also find it "sweet" to sleep when tired, to eat when hungry, to have children, to achieve sexual satisfaction, to master challenges, to communicate, to relate to other human beings, to behave in particular ways in particular circumstances. This is a prominent way genes get their bodies to do something: by endowing those bodies with a natural sweet-tooth that makes it pleasant to do certain things rather than others. Thus do genes prod bodies into doing whatever is good for them (for the genes, that is, although typically good for the bodies too).

The imprint of evolution can sometimes seem a bit heavy-handed, and not always benevolent. In the next few chapters, for example, we'll consider why people sometimes find it sweet to prefer relatives over strangers, why they generally favor biological children over step-children, and so forth. In the process, you may come to see yourself and others around you in a new light, as survival machines put together by genes that have proved their worth through millions of years of evolution. Increasingly, these survival machines and their genes are even coming to understand themselves.

"Altogether, man is a darkened being," wrote Goethe. "He knows not whence he comes, nor whither he goes; he knows little of the world, and least of all himself."[20] This may have been true in Goethe's time, but now, with the dawning of revolutionary biology, it is changing.

Denying Our Nothingness

What is the effect of applying evolutionary biology to *Homo sapiens*? For one thing, by focusing on the genetic underpinnings that all people share, it offers a new way of understanding that elusive quarry, "human nature." Human nature is not like a unicorn or some other mythical beast. It exists. It exists because human genes exist, and because they produce a different kind of creature from hippopotamus genes or hyacinth genes. Notwithstanding all the variety displayed by human beings in different cultures and societies, the fact remains that nowhere are there people who make do without love, marriage, art, competition, language, tools, smiling and laughing, crying and yelling, adornment for their bodies, and care of their young, to mention just a few.

Also, nowhere are there people among whom men are less violent than women, who consider eighty year olds more sexually attractive than twenty year olds, or who do not favor their relatives over strangers. And this, too, is an incomplete list.

That elusive yet powerful undercurrent, human nature, flows like a subterranean stream through the geography of the past and the present. It is the unseen but constant presence that undergirds the universality of great works of creative genius. It is, for example, what makes Shakespeare's characters readily comprehensible, despite the fact that they are now 400 years old. Their language may sometimes be dauntingly archaic, but this only emphasizes the point: Even as the surface features of expressiveness have greatly changed, some things remain stodgily the same. And so, there is something familiar and recognizable about such basic, such obviously *human* traits as Romeo and Juliet's hormonally overheated teen-age love, Hamlet's intellectualized indecisiveness, Lady Macbeth's ambition as well as her guilty remorse, Falstaff's drunken cavorting, Viola's resourcefulness, Lear's impotent rage, Othello's jealousy, and Puck's, well, his Puckishness. And when the latter concludes, wonderingly, in *A Midsummer Night's Dream*, "What fools these mortals be," we cannot but agree, because we know what human beings they—and we—all are.

To those who worry that human beings are somehow diminished by acknowledging their basic, biological, evolutionary underpinnings, I would argue that precisely the opposite is true. A genetic perspective emphasizes that instead of a blank slate or vacant vessel, to be written upon, filled up, or in some other way passively created by their experiences, people *are* something. They matter. "The greatest mystery," according to André Malraux, "is not that we have been flung at random among the profusion of the earth and the galaxy of the stars, but that in this prison we can fashion images of ourselves sufficiently powerful to deny our nothingness."[21] Revolutionary biology makes this definite contribution: It gives us ammunition with which to deny our nothingness. "*This*," it says, "*this* is what a human being *is*."

Notes

1. Richard D. Alexander. 1981. Evolution, culture and human behavior: Some general considerations. In Natural Selection and Social Behavior, D.W. Tinkle and R. D. Alexander, eds. Chiron Press: New York.
2. G. C. Williams. 1966. Adaptation and natural selection: A critique of some current evolutionary thought. Princeton University Press: Princeton, NJ.

3. Sigmund Freud. 1914 (1959). On Narcissism. Basic Books: New York.

4. Daniel C Dennett. 1995. Darwin's Dangerous Idea. Simon & Schuster: New York.

5. For more on the evolutionary biology of male-female differences, see David P. Barash and Judith Eve Lipton. 1997. Making Sense of Sex. Island Press: Washington, D.C.

6. Donald C. Peattie. 1935. An Almanac For Moderns. G. P. Putnam's Sons: New York.

7. Richard Dawkins, 1989. The Selfish Gene. Oxford Unviersity Press: New York.

8. I would modify this as follows: "To be sure, many of them actually did die out, to be replaced by others. Those replicators that currently exist in the plants and animals of the world are the decendants of those replicators that did not die out. And whereas chance undoubtedly determined the fate of many of them, another important factor was how effective they were at creating survival machines that were well built and that behaved appropriately; that is, that were adapted to their environment and more fit than their competitors."

9. John Crabbe et al. 1996. Elevated alcohol consumption in null mutant mice lacking 5-HT1B receptors. Nature Genetics 14: 98-101.

10. David G. Freedman. 1958. Constitutional and environmental interactions in rearing of four breeds of dogs. Science 127: 585-586.

11. Mario de Bono and Cornelia Barghmann. 1998. Nocturnal variation in a neuropeptide Y receptor homologt modifies social behavior and food re sponse in *C. elegans*. Cell 94: 679-689.

12. Nicholas Wade "First gene for social behavior identified in whiskery mice" NY Times, Sept. 9, 1997, B10.

13. Richard P. Ebstein, et al. 1996. Dopamine D4 receptor (D4DR) exon 3 polymorphism associated with the human personality trait of naovelty seeking. Nature Genetics 12: 78-80; Jonathan Benjamin et al. 1996. Population and familial association between the D4 dopamine receptor gene and measures of novelty seeking. Nature Genetics 12: 248-256.

14. Within a year after this research was reported, a different team of researchers was unable to replicate its findings (D. J. Vandenbergh et al. 1996. No association between novelty seeking and dopamine D4 receptor. Molecular Psychiatry 2: 417-419). It is possible that the initial results were in error, that something was flawed in the failure to replicate, or that there are many genes "for" thrill-seeking, so that the one in question is not always correlated with the behavior at issue. Although the jury is out on this particular example, it is almost certain that as genetic research picks up steam, it and/or other examples will be confirmed in the near future.

15. K-P Lesch, D. Bengel, A. Heils, S. Z. Sabol, B. D. Greenberg, S. Petri, J. Benjamin, C. R. Muller, D. H. Hamer and D. L. Murphy. 1996.. Association of anxiety-related traits with a polymorphism in the serotonin transporter gene regulatory system. Science 274: 1527-1531.

16. I can see it now: Someone will accuse me of "really" believing there is a gene for buying Volvos...as distinct from, say, a gene for Corvettes.

17. J. M. Bailey and R. C. Pillard. 1991. A genetic study of male sexual orientation. Archives of General Psychiatry. 48: 1089-1096; J. M. Bailey, R. C. Pillard, M. C. Neale, and Y. Agyei. 1993. Heritable factors influence sexual orientation in women. Archives of General Psychiatry 50: 217-223.

18. Gerald E. McClearn, B. Johansson, S. Berg, N. L. Pedersen, F. Ahern, S. A.

Petrill, & R. Plomin. 1997. Substantial genetic influence on cognitive abillities in twins 80 or more years old. Science 276: 1560-1563.

19. Leibnitz was a renowned 17th century philosopher who maintained that the world was composed of irreducible building blocks, which he called monads; in a sense, his view was the polar opposite of the philosopher Spinoza, who maintained that all things were fundamentally unified.

20. Conversations of Goethe and Echermann. 1829 (1930) trans. John Oxenford. J. M. Dent: London.

21. Andre Malraux. 1989. The Walnut Trees of Altenberg. trans. A. W. Fielding. H. Fertig: New York.

2

Altruism: Theory and Animals

A bumper sticker exhorts "Commit random acts of kindness and senseless acts of beauty." The sentiment is admirable, all the more so because acts of kindness are deplorably rare. And when they do occur, they are unlikely to be random.

For "kindness," read "altruism."

Altruism is one of the great mysteries of life, and its seeming solution, one of the triumphs of modern biology. As triumphs go, this one has been especially important, in two ways. First, it is always nice to solve a puzzle, especially one that has stumped the great minds of the past century. And second, the scientific solution to the question of altruism generated much of the explanatory horsepower that has fueled gene-centered biology, and which in turn has revolutionized our understanding of all living things, human beings included.

The Crux of the Problem

Why was altruism such a mystery? Because it shouldn't exist. After all, evolution works by "differential reproduction," that is, by certain individuals and their genes out-reproducing others. Natural selection, therefore, is downright selfish. Marine biologist and historian of science Michael Ghiselin put it chillingly and well:

> The economy of Nature is competitive from beginning to end.... No hint of genuine charity ameliorates our vision of society, once sentimentalism has been laid aside. What passes for cooperation turns out to be a mixture of opportunism and exploitation. The impulses that lead one animal to sacrifice himself for another

37

turn out to have their ultimate rationale in gaining advantage over a third; and acts "for the good" of one society turn out to be performed to the detriment of the rest. Where it is in his own interest, every organism may reasonably be expected to aid his fellows. Where he has no alternative, he submits to the yoke of communal servitude. Yet given a full chance to act in his own interest, nothing but expediency will restrain him from brutalizing, from maiming, from murdering—his brother, his mate, his parent or his child. Scratch an "altruist," and watch a "hypocrite" bleed.[1]

Evolution is not immoral. But neither is it moral. It is amoral, simply statistical, and one hundred percent indifferent to ethics. Success—not goodness—is rewarded. Failure is the only sin. All that counts is the notorious bottom line, measured not as money or profit, but fitness. And fitness is a matter of numbers, of cold-heartedly assessing genetic profit margins, in which the miserly, evolutionary Scrooge only wants to know how many genetic descendants are catapulted into the future.

The problem is simple: Evolution rewards selfishness. Insofar as a trait or behavior increases reproductive success, that trait or behavior should become more abundant, along with its corresponding gene(s). At the same time, any trait or behavior or gene(s) that reduced reproductive success should quickly disappear to be replaced with its selfish alternatives. So what about altruism?

For example, imagine there were genes for "midwife skills," and any individual carrying such genes was particularly adept at bringing babies into the world. If being a wonderful midwife brought peace and harmony, or money, or status, or something else that made it easier for its owner to have lots of happy, fertile children, then midwife genes would spread. However, if these genetically inspired midwives had fewer successful children, because of job stress, or being underpaid, or simply because attention to everyone else's reproduction took time away from their own, then those genes—admirable and useful as they might be—would die out. An individual with such genes might have ample social success, money, status, and love, but if they interfered with making his or her own babies, then the genes in question would die away.

Adaptation, in short, means looking out for "number one," one's self and one's offspring. Individuals (or their genes) who look out for someone else instead must be at an evolutionary disadvantage, so long as this "looking out" imposes a cost on the looker, and confers some benefit upon the recipient, who, after all, is number two insofar as natural selection measures these things.

Natural selection helps those who help themselves. And it penalizes those who help others. As a result, altruism should disappear, or, if it occasionally pops up because of an errant mutation, it should quickly be selected against and replaced by selfishness—which, by definition, helps itself and thus prospers. Another way of stating the problem: How to explain the endurance of any trait which is, by definition, self-defeating?

And yet, there are cases of undoubted altruism among animals, examples of animals doing things that benefit another individual at some cost to the benefactor. In fact, there are many such cases. Let's consider one example, known to biologists as alarm-calling. Among animals that live in social groups—say, prairie dogs—alarm calls are often given when a predator such as a hawk or coyote appears. Other prairie dogs hearing the alarm respond by running to their burrows. The alarm-caller is an altruist. After all, he or she did something (sounded the alarm) which benefited others—those who heard the alarm. Moreover, there was some cost to the altruist. Bear in mind that when the alarm-caller alerts others that a coyote is in the neighborhood, he or she also alerts the coyote to the presence of the alarm-caller! (Often, people become aware of prairie dogs and other alarm-callers only after one of them sounds the alarm; not surprisingly, alarm-callers are conspicuous.)

Why does the coyote-spotter spread the alarm, instead of quietly and inconspicuously going into its burrow, and letting its fellow prairie dogs suffer the consequences of their ignorance? Why isn't it the strong, silent and selfish type, instead of being a noisy, alarm-calling altruist? It would seem that silent selfishness would be selected for— that is, it would be more fit—while self-denying behavior of this sort would disappear. After all, Darwin himself wrote that evolution "will never produce anything in a being that is injurious to itself, for natural selection acts solely by and for the good of each."[2] Yet alarm-calling, although good for those who hear it, is generally injurious to the alarm-caller.

Darwin also pointed out that "every complex structure and instinct [is likely to be] . . . useful to the possessor." As a result, there shouldn't be any persistent behavior that is useful to another individual, and certainly none that is downright *harmful* to the possessor. Nice guys, in short, should finish last; so much so that eventually, they shouldn't even be able to show up for the next season. Like any team that loses all of its games, altruists should go extinct. And this should apply not only to alarm-calling, but to other tendencies to risk one's life on another's behalf, or even to share food with someone else, and so forth.

But nice guys do rather well. Altruism not only astounds, it abounds. Why?

The Solution

Ironically, one of the earliest obstacles to solving the problem of altruism came from the fact that it was *too* easy to explain, so easy that it wasn't considered much of a problem at all. It had long been assumed that evolution operated via "the good of the species," or "for the benefit of the group." If so, then altruism isn't surprising. Indeed, it is just what you might expect: Sharing food, risking one's life for the benefit of others, scratching someone else's itch . . . such things ought to be the norm if natural selection submerges the individual in a benevolent concern for the greater good of the greater group, a larger and presumably more important entity of which the individual is but a small part. Another way to look at it: Species containing altruists might very well do better than species made up entirely of selfish individuals, and so, if natural selection operates for the good of the species, then *of course* individuals should be altruistic. It's good for the species.

Accordingly, just as altruism is difficult to reconcile with evolution operating on the *individual* or the individual *gene* (because by definition, such individuals lose out when they behave altruistically), altruism is easy to reconcile with evolution operating on groups (because, almost but not quite by definition, groups do better if they are composed of altruists).

There is a huge problem with this reconciliation, however: It is based on illusion.

The difficulty is as follows. Let's use food-sharing as our example this time. Imagine that a group of animals—call them Hannahs—contains a few altruists. These helpful Hannahs give up their food to other, hungry Hannahs. As a result, any Hannah groups containing altruists can be expected to prosper; after all, hungry Hannahs don't starve because they are assisted by their altruistic, helpful associates. We might even expect that Hannah groups with altruists will do better than other groups whose constituents are entirely selfish.

The worm in the center of this apple is that within those groups blessed with altruists, helpful Hannahs will be at a *disadvantage* compared to any individuals that are selfish. When short of food themselves, these hard-hearted Hannahs may well profit from the generosity of their colleagues. Worse yet, they will never suffer any of the costs, which are borne entirely by the helpful Hannahs. As a result, helpfulness among Hannahs should become as scarce as self-destructiveness of any other sort, even though such behavior is actually helping someone else. In fact, it is self-destructive, in part, *because* it is helping someone else.

Altruism is, in a sense, twice costly, since by helping someone else, the altruist is losing out in the process and also in the exchange, placing him or herself at a double disadvantage, both absolute and competitive. After all, natural selection operates by differences in reproductive success: If X does better and Y does worse as a result of some altruistic action by Y, then this also means that X is likely to be doing *better than Y as a result*, which, in a sense adds insult to injury, providing yet a further disadvantage to Y's altruism.

Thus, rather than solving the problem of altruism, explanations that rely on supposed group benefits if anything highlight the discrepancy between altruism and selfishness, and the reason altruism is so perplexing: because altruists do worse than non-altruists, any way you slice it. Or so it seems.

The solution is a kind of Goldilocks Principle: just as explanations in terms of group or species benefit are too easy, and explanations in terms of individual benefit are too difficult, those focusing on the gene are just right.

True to the approach of the first chapter, let's take the perspective of a hypothetical gene for alarm-calling; not an alarm-calling individual, or a group containing alarm-callers, but an alarm-calling *gene*. Designate it gene "a," as opposed to its silent, selfish alternative, gene "s." If "a" induces its body to give alarm-calls, as a result of which that body and its "a" gene end up inside a coyote's belly, then the "a" gene isn't doing very well by itself, or by its survival machine. In such a case, the "a" gene should disappear along with its unfortunate, altruistic body, to be replaced in the population by "s" genes residing in bodies that we identify as selfish.

This gene's-eye view actually contains the answer hidden within it. The problem of altruism is solved when we realize that what appears to be altruism *at the level of bodies* can be plain old-fashioned selfishness *at the level of genes*. When altruistic prairie dogs shrilly announce the presence of a coyote, they are not simply benefiting other individuals at their own expense. Instead, their alarm calls make it more likely that identical copies of the "a" gene, residing in the bodies of other prairie dogs, will be saved. By influencing their bodies to sing out when they spot a coyote, "a" genes are being just as selfish as evolution could desire, making it more likely that identical copies of those "a" genes will be projected into the future, instead of becoming coyote chow. So, the paradox of altruism disappears with the realization that it isn't really a paradox at all, because it isn't really altruism—at least, not for the genes.

The situation is basically the same for our hypothetical Hannahs. Imagine a helpful, food-sharing gene, "h," competing with its alternative, eat-heartily (and selfishly) gene, "e." So long as the food-sharing "h" genes are likely to share their food with Hannahs carrying other "h" genes, then the tendency for Hannahs to be helpful will be selected for. This is true even though to the uninitiated, helpful Hannahs may look more like suckers than like cagey evolutionary strategists.

Later, we'll return to the case of alarm-calling, because it has received a lot of attention from biologists, and also because it serves as a useful metaphor, helping to clarify the meaning of altruism generally. We began with this example in the hope of drawing in the reader, seducing him with the intellectual problem as well as the peculiar elegance of its solution. But in the process, we inexcusably skipped over the scientist whose remarkable insight made it all possible. (And by "it," we mean not only this solution to the problem of altruism, but also much of gene-based thinking more generally.)

That scientist is William D. Hamilton, a brilliant, soft-spoken, seemingly absent-minded evolutionary geneticist from England who shared with a small number of original thinkers the ability to see, in a new way, what the majority take for granted. Most historians agree, for example, that Isaac Newton wasn't really hit on the head by a falling apple. But it took Newton to recognize that "gravity" was something worth identifying and measuring, a phenomenon to be identified. The physical world was not the same after Isaac Newton:

"Nature and nature's laws lay hid in night:
God said, Let Newton be! and all was light.[3]

The biological world has not seemed the same after William D. Hamilton.

Isaac Newton opened physicists' eyes to why things fall; Hamilton, in effect, opened biologists' eyes to why living things behave as they do—even, why they are what they are. Hamilton's now-classic article, "The genetical evolution of social behavior," published in 1964, is, more than any other single piece of research, the intellectual corner-stone of the modern evolutionary revolution. In effect, Hamilton's insight was to recognize that genes promote their success via copies of themselves in other bodies.

Even before Hamilton, biologists had never been troubled, interest-ingly, by the ubiquity of reproduction, even though *at the level of bodies*, breeding is just as altruistic as alarm-calling. After all, repro-duction is costly. It takes time and energy. It involves risk and im-poses penalties on the would-be breeder. (Think of the time and en-ergy spent in courtship, the vulnerability associated with mating, the sheer metabolic cost of constructing a placenta, lactating, defending and provisioning one's offspring, etc.) Reproducing, in short, benefits someone else—the offspring—while it does immediate harm, if any-thing, to the parent.

Yet parental behavior is not normally considered altruistic; making children is not surprising, nor is it in any way counter-intuitive, or against what an evolutionary biologist—or anyone else—might ex-pect. Quite the opposite: Most people take reproduction for granted, and biologists have long considered that successful breeding is central to evolutionary success. For decades, in fact, biologists equated breed-ing with fitness. Reproduction is costly? Of course. But it would be absurd to think that as a result, reproduction would be selected against! What would replace it? All living things are the offspring of parents who successfully reproduced, costs and all. A genetic basis for non-reproduction would have a dim evolutionary future indeed.

But here's the point: Looked at from the level of the genes, the important thing about reproduction is that those genes are packaging copies of themselves into new bodies, and then (in varying ways, depending on the species) trying to promote the success of these new bodies. How? By feeding them, keeping them warm, protecting them, teaching them, taking them to soccer games and the orthodontist, maybe sending them to college and paying their bills. With this new perspective, having babies and then caring for them is seen for what it is: a perfectly good route to evolutionary success. But not the only route. Hamilton's genius came in recognizing that there are other ways for genes to be successful, if not via those bodies that we call offspring, then via other bodies that we call genetic relatives, such as nieces, nephews, cousins, grandchildren, and so on.

The only difference between these more distant relatives and those that we call offspring is that the more distant the relative, the lower the probability that a gene present in any given individual is also present in that relative. This, after all, is what people mean when they talk about a "distant" relative: genetic distance, even though most of us lack the ability to calculate exactly how great the distance and precisely what genetic "distance" really means.

To emphasize this wrinkle in the altruism story, think of those generalized, random "acts of kindness" with which this chapter began. By now, two things should be clear: first, why acts of this sort are so intriguing to evolutionary biologists. And second, why it is so seldom that they are truly "random." If any gene-based tendency for altruism is to profit by benefiting copies of itself in other bodies, then the "kindness" must not be sprayed about indiscriminately. Hamilton pointed out that it must be directed preferentially toward those with a reasonable chance of possessing the gene(s) in question. In fact, the higher the probability of the gene(s), the higher the probability of the behavior.

Hamilton went farther. He showed the conditions necessary for altruism to evolve. To make a long mathematical story short, "Hamilton's rule" is that altruism will be selected for in proportion as (1) the cost to the altruist is low, (2) the benefit to the recipient is high, and (3) the altruist and recipient are closely related. The first condition means that low-cost altruism—for example, taking a small risk for someone else—should be easier, and thus, more frequent, than running extreme risks. The second means that all things being equal, altruists should be more likely to act in proportion as their altruism helps the individual being assisted. And the third condition means that "more closely related = more altruism" and conversely, "less closely related = less altruism." Why? Because the closer the genetic relationship, the higher the probability that any altruism-promoting gene(s) present in the altruist will also be present in the recipient.

The result is a new picture of evolutionary fitness, one that reveals how the net of natural selection is spread more widely than pre-Hamiltonian Darwinists had imagined. Previously, when biologists thought about fitness, they considered only direct reproductive success; its importance has never been in doubt. But breeding is only part of the story. The full tale, known as "inclusive fitness," is more, well, inclusive. It includes not only reproductive success but also any action that increases another's survival and reproduction. But—and here is a very important point—not all "others" are equal, at least insofar as a would-be altruist is concerned. The importance of each "other" (to the altruist) is greater in proportion as he or she is more closely related.

Actually, there are two noteworthy precursors of Hamilton's important insight. Both were brilliant evolutionists—in fact, they were two of the most prominent founders of the field of population genetics—but for some reason, neither carried this particular idea very far. In the late 1920s, Ronald A. Fisher wondered why certain bad-tasting caterpillars were brightly colored. He acknowledged that conspicuous coloration would make it more likely that a hungry bird, for example, after eating one caterpillar, would leave the others alone. But, Fisher pointed out, such an "advantage" would come a bit late for the caterpillar who sacrificed its life in order to educate predators not make the same mistake twice. Fisher went further, suggesting that perhaps this is why such insects tend to be found in groups: If these groups consist of brothers and sisters, then the dying caterpillar would be repaid—not in this life, but in evolutionary time—through the success of numerous kin.

The other biologist who caught a glimpse of the genetics of altruism but apparently did not realize its generalizability was J. B. S. Haldane, like his contemporary Fisher and his successor, Hamilton, a British mathematics whiz. The story goes that Haldane was at his favorite pub when the conversation happened upon self-sacrificial bravery. Haldane was asked if he would give his life for his brother. No, he said, he wouldn't do that. Then he made a rapid calculation on the back of a napkin and added that he'd do so for two brothers or eight cousins!

This, apparently, is as far as it went, until Hamilton revisited the paradox of altruism, bequeathing us a new view of ourselves and of life more generally. The result is also sometimes called "kin selection," since it speaks to a predictable bias toward kin: relatives over non-relatives, and closer relatives over more distant ones. Kin selection—or "inclusive fitness theory"—suggests that nepotism is likely to be universal, or nearly so, in the living world. It even provides a way of calculating it. As J. B. S. Haldane recognized, one self equals two brothers, or four grandchildren, or eight cousins, and so on. Faced with the question, "Save your skin or save your kin?" the balance point occurs when the likelihood of genes present in relatives equal those present in one's self.

Armed with this new view of behavior, biologists have begun to reinterpret the living world.

***

To the reader: The rest of this chapter considers various aspects of inclusive fitness applied to animals. This is because I am fascinated by animals, and excited about the new universe of understanding that kin selection has opened. The examples that follow should also help deepen and broaden your appreciation of the ins and outs of animal altruism. Nonetheless, if you are reading this book because you want to learn about people and nothing but people, you might want to skip ahead to the next chapter. I think you would miss some wonderful stuff, but having read this far, you have what you need in order to make the leap to human beings.

***

Alarm-Calling, Sharing, Fighting

When scientists first got hold of telescopes, and microscopes, they eagerly turned their instruments on everything in sight—or out of sight. The biology of altruism offered another kind of intellectual lens, and, true to form, when biologists got hold of it, they turned their just-discovered conceptual tool on every example of animal social behavior they could find. The results have been gratifying.

It had long been known, for example, that when a zebra is attacked by lions, other zebras may come to the rescue and try—albeit often unsuccessfully—to drive the predators off. By contrast, when a wildebeest is attacked by lions, other wildebeest typically do not get involved. Sure enough, wildlife researchers quickly pointed out that zebras live in extended family groups, whereas wildebeest—especially during migration—travel in a kind of primordial "lonely crowd," in which individuals are generally surrounded by an anonymous horde of nonrelatives. As a result, individual zebras can enhance their inclusive fitness by attempting to defend a relative, whereas for wildebeest, making common cause against a hungry lion is, if not suicidal, unlikely to bring any genetic return.

Standing up to lions is highly altruistic, since it occurs at substantial risk to the altruist, and with large potential gain for the potential victim. Hence, in terms of the underlying biology of altruism, zebra bodies lose out, while zebra altruism genes move ahead. On the other hand, altruistic defense by wildebeest would be a losing proposition all around: Not only would the defending animals lose fitness (since they, like zebras, might be killed or injured in the process), but since the beneficiary is unrelated to the altruist, there would be no compensating gain for the genes involved.

In short, zebra "altruism" is really genetic selfishness, and it happens. Wildebeest "altruism" would be genuine altruism indeed—at the level of genes no less than of bodies—and not surprisingly, it doesn't happen.

Return now to our initial example of altruism, alarm-calling. For it to be favored by kin selection, alarm-calling would have to be directed preferentially toward genetic relatives. Otherwise, if non-callers and callers alike were equally likely to benefit from alarm calling, the non-callers and their genes would come out ahead since they would enjoy the same benefits as their more generous and vocal colleagues, without suffering any costs. Paul Sherman of Cornell University therefore predicted that alarm-calling ground squirrels should be those that had genetic relatives above ground when a predator appeared. Prediction confirmed. Furthermore, since females remain as adults in the colony into which they are born whereas males emigrate to new social surroundings, females should generally be more prone than males to give alarm calls; they would be more likely to have relatives among those likely to profit from their calling, while newly arrived males would be surrounded by comparative strangers. Confirmed again.

Sherman also predicted that there should be less fighting among relatives than among non-relatives, and less fighting among close relatives than among distant ones. Once more, the answer was yes. Close relatives also turned out to be especially likely to cooperate in defending their territories. In addition, relatives were more prone to share burrows than non-relatives, and close relatives more so than distant relatives.[4]

What about primates? Eager to test the power of kin-selection on our closer relatives, anthropologist Jeffrey Kurland of Penn State University studied the behavior of Japanese macaque monkeys, also looking at affiliative versus aggressive interactions. As expected, he found that relatives behaved more affiliatively (more social grooming and other positive, "friendly" activities) than non-relatives. But—surprise!—closely related Japanese macaques also turned out to be *more* aggressive toward each other. Some reflection suggests the explanation: Even human beings, who consistently tend to favor their relatives, are also more likely to argue, say, with family members than with strangers, if only because they spend so much more time with the former. (Another question that presents—and answers—itself is *"Why* do people spend more time with family members?" Mightn't that have something to do with shared genes as well?)

When Kurland re-analyzed his results to account for the greater physical proximity of genetic relatives, he found that compared to non-relatives, family members among Japanese macaques were, in fact, much *less* likely to behave aggressively toward each other than were unrelated individuals.[5]

This pattern of reduced aggression among close relatives has been searched for in many different animal species, and nearly always, it has been found.

Here is another example, involving another species of macaques, rhesus monkeys living in semi-wild conditions on a small island near Puerto Rico where they had been introduced. (This is the species, incidentally, in which the blood gene known as "rh-factor"—for "*rh*esus"—was first discovered.) Since births had long been carefully recorded among these closely studied animals, it was possible to calculate the average degree of genetic relatedness within a rhesus troop. As time went on and the population increased, genetic closeness decreased. This makes sense: Imagine a small group consisting of a few families, mostly parents and their offspring. Eventually, the founding parents die, and after them the first generation of siblings, who are replaced by aunts, uncles, nieces, nephews, first cousins, then second cousins, and so on.

As the genetic relatedness among these monkeys declined, aggressive incidents went up, until eventually, these ever larger, ever more unruly, ever more unrelated groups split up into smaller groups, each of which was smaller, better behaved, and more closely related. Thus, when these episodes of "group fissioning" took place, relatives tended to stick together, which resulted in a high average genetic relatedness; moreover, within these groups of close relatives, aggression and conflict returned to its earlier, low level.[6] Once again, a familiar pattern was revealed: Close relatives act more benevolently—that is, more altruistically—toward each other.

It is a pattern that extends in a number of directions, some of them surprising.

Mating Systems

The sexual antics of wild turkeys are a bit perplexing. At first blush, courting turkeys don't seem at all altruistic, but in this case, looks are deceiving. To begin with, male turkeys face off in groups of two. Within each little duo, competition is fierce, with rampant displaying and occasional fighting, as each individual tries to dominate its "partner." Eventually, things calm down as a simple rank-order is established in each pair: one is dominant; the other, subordinate. Then, displaying and fighting flares up once again, except this time pairs compete with other pairs. Each pair now functions as a tight little social unit, cooperating in combat against other, comparable duos. Within each twosome, subordinates back up their former rival, who has become the dominant member.[7]

Eventually, a top-ranking turkey twosome emerges, whereupon the goal of all this bickering becomes clear: Turkey hens make their appearance, and they proceed to mate, not at random, but with the dominant male in the dominant twosome. This system makes perfect sense, for the turkey hens at least. By mating with the dominant male of the dominant pair, female turkeys are getting the best genes around: the crème de la crème. Similarly, we can easily understand the benefit derived by the dominant male: He gets to perpetuate his genes via many willing turkey hens. (Translated into gene language: Genes that succeed in creating a dominant male get to project copies of themselves via large numbers of willing turkey females.)

This presumably, is the pay-off that drives the male willingness to indulge in so much time-consuming and potentially dangerous fighting. But what about the subordinate male? What does he get out of the arrangement? Indeed, the same question arises for any of the "beta" turkeys: After failing in his initial contest for supremacy within each twosome, why does the subordinate continue to participate in the social melee, supporting the dominant alpha male in the ensuing competition among twosomes? Remember, even if the duo is successful, it is the dominant male who will get to do nearly all the mating. What is going on here? Is the subordinate really such a turkey?

As it turns out, there is selfish, genetic method behind his seemingly altruistic madness. Turkey twosomes are usually composed of brothers, and so, they are close relatives; full siblings share a 50 percent probability that a gene possessed by one sibling is present in the other because of their common parentage. From the evolutionary perspective of the subordinate male, therefore, even if he is unable to become a father, he has a chance of being an uncle, thereby promoting his inclusive fitness by proxy, through the children that will be sired by his brother.

The North American turkeys may well exemplify a widespread evolutionary strategy: first, try to enhance your evolutionary success in the most direct way possible, by successful mating. If this requires competing with others, then go for it. But at the same time, keep your tactical options open for a fall-back position. If you can't get your first choice—parenthood—then settle for the next-best thing: in the case of the turkeys, unclehood. (It is interesting to consider that Benjamin Franklin, a wily old codger himself, once proposed the turkey as the U.S. national symbol.)

Altruistic mating cooperation of the turkey sort is especially striking when it is performed by males, since the typical pattern for most species is for males to compete—often violently—with each other, attempting to copulate with as many females as possible. It is also found among some mammals, including lions. Male lions often form mating coalitions which fight it out with other coalitions for control of a pride of females. Not surprisingly, the top male then gets the lion's share of the breeding. As in the case of the wild turkeys, we must ask: What's in it for the subordinate lions, who expend time and energy, and undergo risk of injury, in order for the local lion king to hog most of the reproduction? Again, as with the wild turkeys, the answer is that the various lion princes and assorted gentry get something out of the deal: a distinct, measurable inclusive fitness benefit, because they tend to be closely related to the dominant, breeding male.[8]

There is a twist to this story. Coalitions consisting of more male lions are generally successful in ousting smaller coalitions; this would seem to suggest that lions should join larger coalitions whenever possible. But here's the catch: The larger the coalition, the lower the chance that a subordinate lion will get any direct matings at all. In a sense, then, subordinate lions who join a large coalition are being more altruistic than those joining a small one, since larger coalitions confer a greater benefit on the leader while imposing a greater cost on the subordinate member of the coalition. Interestingly, Craig Packer of the University of Minnesota and his associates found that large lion coalitions tend to be made up of closer relatives than small coalitions. The evolutionary logic appears to be as follows: For a male lion to join a small pride, the prospect of copping the occasional copulation is sufficient. But to join a large one—in which such possibilities are slim—there must be the alternative possibility of gaining an indirect benefit via the success of relatives.[9]

It appears, therefore, that altruism—operating via kin selection—shows itself not only in direct behaviors such as alarm-calling, and in who-is-nice-to-whom, but also in reproductive maneuvering of the turkey and lion kind. One of the more notable examples of the latter is known as the "helper at the nest" phenomenon, well described among many species of birds.

Here is the basic observation: Often, a young animal—fully adult and physiologically capable of breeding—doesn't do so. Instead, he (usually it is a "he") hangs around the nest of another breeding pair, as a "helper." Such helpers generally perform useful chores. They may contribute to building the nest, assist in bringing food to the nestlings, baby-sit, give alarm calls if a predator shows up and even help defend the nest at genuine risk to their lives. These altruistic acts seem doubly perplexing since not only are the helpers giving of themselves and assisting others, they are also forgoing their own reproduction.

The mystery is largely dispelled, however, when we add this crucial bit of information: Helpers are typically offspring of the breeding pair, aiding their parents to rear siblings for themselves. If they were helping total strangers, the situation would be much more difficult to understand, but as it is, the helper at the nest—like the subordinate turkey or the celibate lion—is gaining inclusive fitness by assisting a relative to reproduce.[10] In the lion and turkey cases, we see individuals cooperating with siblings; the standard pattern for helpers at the nest involves individuals cooperating with their parents to help rear additional siblings. The details are different for each species, but the basic pattern is convincing, as once again, apparent altruism is unmasked as genetic selfishness.

The Strange and Special Case of the Bees, Wasps, and Ants

For yet another example of kin selected altruism among animals, we turn to one that is especially convincing. It was also highly influential in Hamilton's thinking: the remarkable case of the "eusocial Hymenoptera." This polysyllabic mouthful occupies a place of honor in revolutionary biology, and it yields such elegant support for gene-centered thinking that it deserves our attention, even though it may be a bit daunting for persons untrained in genetics.

Eusocial (pronounced "you-social") means "perfectly social," and eusocial animals are those enjoying a degree of social cooperation that is, if not perfect, at least remarkable. Another way of looking at it: They are overwhelmingly altruistic, especially with regard to their breeding. Harvard biologist Edward O. Wilson pointed out the counterintuitive reality that sex is actually a disruptive force in animal social life, since individuals often compete with each other, attempting to breed at another's expense. Among eusocial species, however, there is very little sign of sexual competition. Instead, members of the social unit cooperate in their reproductive efforts, combining forces and pulling together for what appears to be the common good, namely, the reproductive success of a small number of breeders within each hive or colony. Foremost among these breeders is the "queen" whose royal prerogative it is to reproduce: She does it, the workers don't. To breed is human, and also animal, as well as plant. To forego breeding is, if not divine, then almost supernaturally altruistic. (Even the avian helpers at the nest may eventually get to reproduce, as we shall see.) But total, lifelong, altruistic nonbreeding is commonplace among the eusocial Hymenoptera.

Such paragons of altruistic cooperation are known to have evolved twelve (possibly thirteen) times independently among animals. Of these, eleven occurred among one group of insects, the Hymenoptera, which includes the ants, bees, and wasps. Among these animals, workers—all of whom are female—do not reproduce. Instead, they labor long and hard on behalf of another individual, the queen. Best known, perhaps, is that when their hive is threatened, many Hymenopterans engage in dramatic, suicidal defense, stinging intruders even though it often results in their death. (The stingers of many bees are barbed, so when they remain in the flesh of a victim, crucial parts of the bee's abdomen are also left behind.)

Here's another example of outlandishly self-sacrificial behavior, less familiar than stinging bees but even more dramatic. In one species of ant, there is a specialized caste that is outfitted with an oversized abdominal gland filled to near-bursting with sticky, resinous, gluey glop. Should an intruder appear, these chemical warfare specialists rush to battle, all the while violently constricting their abdominal muscles, which explode the gland, killing themselves, but spraying the disabling substance all over their enemies.[11]

The really extraordinary thing about stinging bees or exploding ants, however, is not their kamikaze behavior, but the fact that they are permanently sterile. Given that they are sterile, such self-sacrificial behavior in defense of the hive is not really so surprising. After all, animals who have already foregone reproduction don't give up very much more by losing their lives: Once these workers or soldiers are nonreproductive, they are already, in a sense, dead. The big question is: Why are they nonreproductive in the first place?

The remarkable fact, then, is the *sterility* of worker bees, wasps, and ants. The fact that sterility is so concentrated in one particular group of living things suggests that there might be something special about the Hymenoptera that predisposes them to such an altruistic overdose.

There is.

Most animals—including human beings—are "diploid," meaning they carry two sets of chromosomes and thus, two sets of genes. In order to prevent the number of genes from doubling each generation, diploid animals bequeath only one-half their genetic material to each sperm or egg that they produce. These gametes are therefore "haploid," so when two such gametes combine to make a new individual, the normal diploid number is restored. But the ants, bees, and wasps do things differently. Although Hymenopteran females are diploid, just like the rest of us, the males are haploid, like an egg or sperm cell. How is this accomplished? Whereas female ants, bees, and wasps are produced by the union of eggs and sperm, males come into being when some of the queen's eggs develop by parthenogenesis, without being fertilized. The result is "haplo-diploidy," with females diploid (containing a double set of genes) and males haploid (a single set). Therefore, even though a Hymenopteran male looks normal enough, it has only been dealt half a deck of genetic cards. This peculiar limitation has led to some equally peculiar consequences.

Remember that in order to reproduce, a diploid animal engages in a "reduction division" in the course of making its gametes, thereby producing haploid sex cells. Female Hymenopterans do just this when producing eggs. By contrast, male Hymenopterans, as we have just seen, are already haploid, and so, they do not undergo any further reduction division. Instead, their sperm are all genetically identical to each other, and also genetically identical to those of the male producing them. This plays havoc with the usual genetic symmetries of most other living things.

First, we'll give you the results. Then, we'll explain why they occur. And finally, we'll return to Hamilton's explanation, and show how this outlandish genetic system perfectly and elegantly explains the equally outlandish Hymenopteran penchant for eusociality.

The results. Among all mammals, including human beings, parents have a genetic correlation of 1/2 with their offspring; in other words, a gene present in a parent has a 1/2 probability of also being present in any given offspring. Similarly, full siblings have a genetic correlation of 1/2 with each other. By contrast, among the Hymenoptera, full sibling sisters have a higher genetic correlation—3/4—with each other.

The reason. Because their father was already haploid (and didn't undergo a reduction division when producing his sperm), every Hymenopteran sister received the exact same genetic material from her father. Any two such sisters are therefore genetically identical with regard to 50 percent of their genes. So, without even considering any genetic similarity via their mother, these sisters already have a genetic similarity of 1/2, just because they have the same, haploid father. Turning now to the rest of their genes, contributed by their mother, we find that sisters share an additional genetic similarity of 1/2 of that remaining 50 percent, because their mother produced eggs by normal reduction division. As a result, sisters share a total genetic identity of 1/2 (from their father) plus an additional 1/4 (1/2 of the remaining 50 percent, from their mother), for a grand total of 3/4.

The consequences. Please don't miss the forest (in this case, the extreme form of altruism known as eusociality) for the trees: the pecu-

liar genetic system known as haplo-diploidy. Remen
been to explain how the former derived from the la
explanation: A worker bee has, in a sense, a choice. S
to reproduce, like most other animals do. Or she coul
help rear her sisters, additional offspring of her mothe ، ...ے queen. If our worker reproduced, she would have a genetic correlation of 1/2 with each of her children. But if she stayed home, tended the hive, defended it against predators, went on regular foraging excursions, etc., she would be helping her sisters, with whom she has a genetic correlation of 3/4. In other words, from the perspective of her genes, a worker bee does more to enhance her total, inclusive fitness by staying home and being nonreproductive than if she were to raise her own family.[12]

It is striking, and persuasive, that only female Hymenoptera enjoy this remarkable genetic connection of 3/4 with their sisters, and that all worker Hymenoptera are in fact female. It is additionally striking, and persuasive, that the extreme, nonreproductive altruism of worker Hymenoptera toward their sisters has its mirror image in extreme self-ishness on the part of Hymenoptera drones, all of whom are male. Male Hymenoptera, once again, are haploid, and so unlike worker females, which have both father and mother, they are fatherless, pro-duced from unfertilized eggs. These males, being composed of only one-half their mother's genotype (and with no genetic bequethal from a father), have a genetic correlation with their siblings of only 1/4. Rather than be altruistic, they are indeed "lazy as a drone." In short, selfish.

Biologists Robert Trivers and Hope Hare looked further into the tangled web of Hymenoptera genetics, asking whether the behavior of workers reflected their evolutionary interests. Recall, for example, that a tendency for altruistic alarm-calling can evolve only if alarm-callers directed their altruism toward other alarm-calling genes, residing within their relatives. Trivers and Hare reasoned similarly that worker ants should direct their labors toward relatives in proportion to their relat-edness. Since workers share a genetic correlation of 3/4 with their sisters, and only 1/4 with their brothers, they could be expected to feed their sisters, for example, three times as much or as readily, as their brothers. Examining a large body of data, the researchers found, as predicted, that the results did not differ statistically from the predicted three to one ratio.[13]

To wrap up this part of the altruism puzzle: Considering those mals with the most extreme and unusual form of altruism, we find that they have an equally extreme and unusual genetic system, and that the two are precisely consistent with each other. Small wonder that kin selection and the theory of inclusive fitness has taken evolutionary biology by storm.

Improved Visions and New Horizons

Imagine that you had been seriously nearsighted all your life and then, suddenly, you put on eyeglasses. What had been blurry is now clear; many other things, not previously seen at all, are glimpsed at last. This catches some of the effect of kin selection and inclusive fitness theory on those who had been myopically struggling for a good look at the living world.

Some of these new visions were small, precise, and particular, limited in scope but nonetheless fascinating in their own limited sphere. Others were breathtaking in their implications, opening whole new horizons of thought and interpretation.

First, one of the small ones. Shigeyuki Aoki, a Japanese biologist, discovered that in a species of aphid, two different kinds of larvae hatch initially out of their eggs. The normal, "primary" form undergoes several additional molts until becoming an adult, which, in turn, eventually reproduces. The other, "secondary" larval form, is a reproductive dead end. It never molts, never becomes an adult, and thus, cannot breed. Primary aphid larvae have long, slender mouth tubes, adapted for sucking plant juices. The secondary forms have powerful, enlarged forelegs and short, sturdy mouthparts, nearly useless for sucking sap, but very effective in piercing the bodies of any predators who attack the primary larvae.[14]

The secondary larval forms are aphid soldiers, and they are self-sacrificing in the extreme. All that they can do is die for their primary "sisters." Here is the key observation: Both the primary and secondary aphid larvae are the products of *asexual* reproduction on the part of their mother. Accordingly, they carry the unusually high genetic correlation of Hymenoptera sisters one step further: They are genetically identical. As a result, the surprisingly intense altruism of the secondary forms makes perfect sense, in that they aren't really sacrificing themselves at all! (At least, their genes aren't.) Since secondary aphids are genetically identical to the primaries, every time a primary succeeds, the secondaries succeed too. Another way of looking at it:

There are two types of bodies, but the underlying genes are the same and so, there is no conflict of interest between them. The secondary soldiers are like the muscles of a single body, while the primaries are like the ovaries.

This leads directly to the next vision, which goes beyond the admittedly narrow world of aphids to encompass most of life.

Think about any multicellular body, say, a bird, horse, or human being. Now, ask yourself why the liver or kidney cells, for example, willingly undergo the unpleasant task of detoxifying the blood, leaving the fun part—reproduction—to the gonads. Wearing our inclusive fitness/gene selection eyeglasses, the answer becomes clear: The different parts of a multicellular body cooperate so nicely because at the genetic level, *they aren't different at all*. Like primary and secondary aphids, body cells go the eusocial Hymenoptera one step better, sharing a genetic correlation of 100 percent. This is why bodies stick together. Not surprisingly, they do so even more intensely and cooperatively than the members of a beehive.

Now that we have a new way of looking at the make-up of bodies themselves, we can even speculate about how and why these bodies begin to fall apart. Aging, for instance: As bodies get older, their cells become more and more separated from each other, as each cell line undergoes regular divisions, the process known as mitosis. This separation is less one of space than of time, because as time passes, there is more opportunity for minor copying errors—so-called somatic mutations—to take place and accumulate, interposing themselves between cells that, when younger, had been identical. Could this help explain why old age is a time of increasing weakness, breakdown, and troubles of all sorts? Because as bodies grow older, they consist of cells that grow increasingly unruly and uncooperative as they become more different from each other, and thus, less invested in one another's success.

Furthermore, could a similar process help explain certain forms of cancer? After all, cancers are produced by unrestrained cell division, the "malignancy" of cancer cells being due to the fact that they have gone haywire, reproducing themselves excessively and to the detriment of the rest of the body. Normal, noncancerous cells are a bit like worker bees or secondary aphids: altruistic, in a sense, because they restrain any inclination they might have to reproduce on their own. (As just described, such restraint could as well be considered selfish, since normal cells profit by the reproduction of other body cells to which they are identical.) Now, add the fact that many cancers are produced by mutagenic agents, such as chemicals or radiation that induce genetic changes. One reason for the malignancy of such cancers may be, therefore, that because of these changes, their selfish benefit leads them to be less cooperative than they had been before, when they shared a greater genetic interest in the success of the body as a whole. Is cancer, then, a case of selfish, genetic anarchy?

A Few Messy Details

In this chapter, we explored the problem of altruism, and its solution at the level of (r)evolutionary genetics. Doing so, we have been concerned to make the case for kin selection and inclusive fitness theory, at least as they apply to animals. (In the next chapter, we turn to human beings.) Like most important interpretations of the living world, however, there are complications, wrinkles in what purists might prefer to see as a perfectly smooth theoretical fabric. Thus far, we have focused only on supportive findings, unsullied by the messy equivocations that always seem to arise whenever one looks long and hard at the living world, even through the best of lenses. In the interest of intellectual honesty, however, and also hoping to give the reader a more sophisticated grasp of life's complexity, we turn briefly now to some of the messy details.

This is in fact precisely what biologists did. Within a decade after Hamilton's insights were made known, evolution-minded students of social behavior—soon to be called sociobiologists—felt that they had been introduced to altruism's Rosetta Stone, as long-standing mysteries were solved and factual support for kin selection and inclusive fitness theory poured in. After a while, however, a kind of pendulum effect began, and biologists came to appreciate that altruism is neither as simple, nor as simply explained, as their initial enthusiasm had suggested.

Here is a taste of that newly identified complexity.

Alarm-callers aren't always altruists. This alliterative observation is suggested by the fact that before giving an alarm call, prairie dogs who have spotted a predator will typically position themselves so that they can quickly enter their burrow. So, their actual risk may be low. Moreover, it is even possible that on occasion, alarm-callers are selfishly manipulating the behavior of the "beneficiaries," who become more visible when they flee in panic, thereby drawing the attention of predators and leaving the alarm-caller safer than ever. (With friends like these, who needs enemies?) On the other hand, perhaps an alarm-caller is indeed trying to spoil a predator's hunting, by saying, in effect, "I see you, so you may as well try hunting somewhere else." In this case, the alarm-caller would not be acting out of concern for the alarm-caller's relatives but because a discouraged predator might elect to hunt elsewhere in the future, thus benefiting the caller directly.

Then there are the meerkats, a species of African mongoose sometimes called "suricates" and which look a bit like elongated skinny prairie dogs. Here, sentinels remain conspicuously on guard. During a study involving more than 2,000 hours of observation, no raised meerkat sentinel was seen to be taken by a predator, perhaps because such individuals are usually the first to detect a predator, and are nearly always positioned close to an escape burrow.[15] In addition, sentinels were especially likely to do their guard-duty after having been well fed, suggesting that it might actually be a *low*-risk behavior that comfortable meerkats do when they don't need to concern themselves with the important (and dangerous) job of getting food. On the other hand, if being a sentinel is actually selfish, why doesn't everyone get in on the act, instead of just one or two at a time, and—most important— why do they bother giving alarm calls when a predator is spotted? The epitome of selfishness would be to keep silent. Nonetheless, such findings underscore the need for additional research; the final word on altruistic alarm-calling and selfish sentinels is not yet in.

There are other situations in which behavior that was initially seen as altruistic may, on closer inspection, be revealed to be potentially selfish after all—not just selfish at the level of genes—which, after all, is the point of inclusive fitness theory—but at the level of the individual. That is, it might directly increase the so-called altruist's personal, old-fashioned, reproductive, Darwinian fitness. The best example of selfishness masquerading as altruism comes from long-term studies by Glenn Wolfenden of the University of South Florida, and his colleagues. Wolfenden has devoted decades to researching the behavior of the Florida scrub jay, a typical "helper at the nest" species. He found that male scrub jay helpers don't merely derive a kin selected benefit from their helping: By hanging around their parents' nest, helpers increase the likelihood that eventually, they will be able to establish their own breeding territory, adjacent to the family homestead. They may even inherit the family property altogether.[16] In the meanwhile, Florida scrub jays gain an inclusive fitness benefit during their apprenticeship, even while they are looking ahead to the possibility of inheriting a bonanza of direct, reproductive pay-off when their aged parents die.

Selfishness—or, more likely, a tactical mixture of selfishness and kin selected altruism—may be more widespread than biologists' initial enthusiasm for kin selection had anticipated. For an extreme example, Israeli zoologist Amotz Zahavi has made the provocative suggestion that *all* seemingly altruistic behavior, including behavior readily attributed to kin selection, actually functions as a kind of self-glorification. According to Zahavi, an "altruist" is in fact a show-off, intentionally subjecting him or herself to a handicap, in order to demonstrate what superior stuff he or she is made of. This might mean either "Look how generous I can be, giving away food! Just think how nice it would be for you to be my mate!" or "Look how superior I must be, since I am able to run such risks or endure such costs, and still thrive!" Either is a far cry from the lay-person's perception of "altruism"— and from the biologist's, too.[17]

Sometimes, selfishness appears as cowardice, which is not surprising, since its inverse, altruism, often shows itself as self-sacrificial courage. Even the "noble" lion is not immune. Thus, female lions in a pride have been observed to hold back from joining in defense of their group against other lion prides. Such reluctant defenders are being selfish in that they profit from being part of the group but apparently don't take sufficient personal pride in their pride to risk themselves in defense of others. This non-involvement is actually consistent with kin selection, since such uninvolved lionesses are generally unrelated to other group members. For the same reason, however, it isn't clear why these selfish shirkers are tolerated by the others.[18]

Even eusociality, crown jewel of evolutionary altruism, is a bit tarnished. As already described, in eleven of the twelve times that eusociality evolved in insects, it was associated with the genetic oddity known as haplo-diploidy. What about the twelfth case; namely, the termites? They are every bit as eusocial as bees, wasps, or ants. Although they look more-or-less Hymenopteran (and are even widely called "white ants"), termites are quite different creatures, as fully diploid as any human being. There is even a somewhat disputed thirteenth case, a bizarre little beast from Ethiopia known as the naked mole-rat, which is almost—if not entirely—eusocial. These small, hair-less, and highly communal creatures, resembling nothing so much as saber-toothed sausages, are also fully diploid (and, interestingly, their life-style is notably termite-like). But they are mammals! Naked mole rats live in underground colonies of seventy to eighty individuals, where they construct elaborate, branching tunnels because of the co-operative earth-moving activities of most of the colony residents. These individuals—both males and females—function as members of a mole-rat chain gang, doing their various jobs—which notably do *not* include reproducing. Breeding is the prerogative of a single large "queen" and several "kings" who reside in royal, reproductive splendor in a central nest chamber.[19]

If this weren't confusing enough, it turns out that there are a num-ber of Hymenopteran species—notably bees and wasps—that are rela-tively solitary, which is about as far as you can get from being eusocial. So, it is possible to be eusocial without being haplo-diploid (termites, naked mole-rats), and haplo-diploid without being eusocial (solitary bees and wasps). Yet another complication: Among many of those Hymenoptera that *are* eusocial, queens may mate with more than one male. This would drastically reduce the genetic relatedness of the workers.

Finally, there are circumstances in which animals perform seem-ingly altruistic behavior toward others who are *not* relatives. Some-times, they don't even belong to the same species. Although genes are almost certainly involved in such cases, these genes are not shared, as in the case of kin selection. This phenomenon, known as reciprocity, is so important—at least for human beings—that we devote all of chapter 4 to it.

There are many other caveats and complications, situations that give pause and demand a more nuanced understanding of inclusive fitness and its actions. The bottom line: Kin selection is not the sole explanation of everything that appears altruistic. It's a shame if this seems to muddy the water. But in fact, the water of life just isn't perfectly clear, and probably never has been, at least not since living things first took form in the murky liquid of the primeval, organic-gumbo pea-soup. In particular, it seems likely that nothing so compli-cated as social interactions will be explained by any single theory, no matter how elegant or impressive. (Even in animals, never mind in people.) The goal of this book is *not* to make a case for kin selection or inclusive fitness theory, and certainly not to argue that such gene-centeredness is *the* key to understanding behavior. Rather, we hope to show how these and other ideas, limitations and all, shed new light.

Although interest in altruism and kin selection has focused on behavior that is obviously directed toward benefiting another individual (and his or her genes), two other points are worth mentioning. First, it isn't strictly necessary that the beneficiary be a genetic relative. All that is required is that individuals having a given gene display fitness-enhancing behavior toward other individuals that are likely to have the same gene. Relatives fit the bill (since relatives by definition are gene-sharers) but in theory at least, non-relatives would do just fine, if only there were some way of knowing that these non-relatives were also carriers of the gene(s) in question. And second, the behavior doesn't strictly have to be the sort that would obviously qualify as altruistic.

One of the problems of studying altruism, and even more so, of conveying the concept to non-specialists is confusion arising from the English word itself. Thus, in normal usage, altruism is taken to mean some sort of conscious decision to help someone else at one's own expense. For evolutionary biologists, by contrast, it is the outcome that matters, not the motivation, or the outward appearance. And so, we can speak of altruistic bees, aphids, liver cells, even viruses on occasion. In theory, behavior far removed from obvious self-sacrifice could also qualify. As Richard Dawkins pointed out,

> Consider a pride of lions gnawing at a kill. An individual who eats less than her physiological requirements is, in effect behaving altruistically towards others who get more as a result. If these others were close kin, such restraint might be favoured by kin selection. But the kind of mutation that could lead to such altruistic restraint could be ludicrously simple. A genetic propensity to bad teeth might slow down the rate at which an individual could chew at the meat. The gene for bad teeth would be, in the full sense of the technical term, a gene for altruism, and it might indeed be favoured by kin selection.[20]

So: perhaps tooth decay is altruism in disguise, maybe alarm-callers aren't really altruistic after all, and it is at least possible that the nifty genetics of haplodiploidy in the eusocial insects isn't all that its cracked up to be.

But keep your eye on the ball. Don't be confused by these and other equivocations in the ongoing search for the biology of altruism. On balance, gene-centered thinking is an exciting and productive way to look at the living world, as good in its own way as the telescope or the microscope. Or, to change the metaphor: The light shed by kin selection, inclusive fitness theory and the overall biology of altruism is bright indeed, although it is composed of different hues, constantly shifting as our understanding grows. Although it doesn't illuminate every nook and cranny, it enables us, at last, to wander about in the biological world without bumping into things.

This chapter began with a true rarity—random acts of kindness—in the process ignoring the rest of the exhortation: acts of beauty. This is not because beauty is unimportant. To the contrary, Keats was correct to announce that beauty is truth and truth, beauty. (Mr. Keats went on to point out, "that is all ye know on earth, and all ye need to know." I'm not so sure about that.) I am confident, however, that a gene-centered view of behavior is both elegant and beautiful, not the least because it does something extraordinary: It helps us see the truth.

Notes

1. 1974. The Economy of Nature and the Evolution of Sex. University of California Press: Berkeley.
2. C. Darwin. 1967. On the Origin of Species by means of Natural Selection or the Preservation of Favoured Races in the Struggle for Life. Atheneum: New York.
3. Quoted by Isaiah Berlin, 1956, Mentor: New York.
4. P. W. Sherman. 1977. Nepotism and the evolution of alarm calls. Science 197: 1246-1253; P. W. Sherman. 1980. The limits of ground squirrel nepotism. In Sociobiology: Beyond nature/nurture? G. Barlow and J. Silverberg, eds. Westview: Boulder, CO.
5. J. Kurland. 1977. Kin selection in the Japanese monkey. Contributions to Primatology, vol. 35. Basel: Karger.
6. B. Chepko-Sade and T. Oliver. 1979. Coefficient of genetic relationship and the probability of intrageneological fission in Macaca mulatta. Behavioral Ecology and Sociobiology 5: 263-278.
7. C. R. Watts and A. W. Stokes. 1971. The social order of turkeys. Scientific American 224: 112-118.
8. B. C. R. Bertram. 1976. Kin selection in lions and evolution. In P. P. G. Bateson and R. A. Hinde, eds. Growing Points in Ethology. Cambridge University Press: Cambridge, U.K., 281-301.
9. C. Packer, D. Schell, and A. E. Pusey. 1990. Why lions form groups: Food is not enough. The American Naturalist 136: 1-19, and C. Packer, D. A. Gilbert, A. E. Pusey, and S. J. O'Brien. 1991. A molecular genetic analysis of kinship and cooperation in African lions. Nature 351: 562-565.
10. P. B. Stacey and W. D. Koenig, eds. 1990. Cooperative Breeding in Birds. Cambridge University Press: Cambridge, U.K.
11. U. Maschwitz, and E. Maschwitz. 1974. Platzende Arbeiterinnen: Eine neue Art der Feindabwehr bei sozialen Hautfluglern. Oecologia 14: 289-294.
12. The key here, from the worker's viewpoint, is not simply that she enjoys a high genetic correlation with her sisters generally, but that some of those sisters will themselves grow up to be queens, and when those queens reproduce, the workers

will profit, via kin selection. In the same sense, of course, the evolutionary importance of children is not the children themselves, but the fact that some of them will reproduce, yielding grandchildren, and so on, ad infinitum.

13. R. L. Trivers and J. Hare. 1976. Haplodiploidy and the evolution of the social insects. Science 179: 90-92.

14. S. Aoki. 1977. *Colophina clematic* (Homoptera, pemphigidae) an aphid species with "soldiers" Kontyu 45: 276-282.

15. T. H. Clutton-Brock, M. J. O'Riain, P. N. M. Brotherton, D. Gaynor, R. Kansky, A. S. Griffin and M. Manser. 1999. Selfish sentinels in cooperative mammals. Science 284: 1640-1644.

16. G. E. Wolfenden and J. W. Fitzpatrick. 1990. Florida scrub jays: A synopsis after 18 years of study. In P. B. Stacey and W. D. Koenig, eds. Coperative Breeding in Birds. Cambridge University Press: Cambridge, U.K., 241-266.

17. A. Zahavi. 1995. Altruism as a handicap—the limitations of kin selection and reciprocity. Journal of Avian Biology 26: 1-3.

18. Robert Heinsohn and Craig Packer. 1995. Complex cooperative strategies in group-territorial African lions. Science 269: 1260-1262.

19. P. W. Sherman, U. M. Jarvis, and R. D. Alexander, eds. 1991 The Biology of the Naked Mole-Rat. Princeton University Press: Princeton, NJ.

20. Richard Dawkins. 1979. Twelve misunderstandings of kin selection. Zeitschrift fur Tierpsychologie 51: 184-200, 190.

3

Human Altruism

"How do I love thee? Let me count thy genes."

It isn't quite so simple. Nonetheless, love and altruism are closely allied: We are likely to behave altruistically toward those we love. Or perhaps "love" is a way of labeling those toward whom we are disposed to be altruistic—even, on occasion, individuals to whom we are *not* genetically related. Since altruism and shared genes are also connected, however, this perversion of Elizabeth Barrett Browning's famous poem may at least capture a part of the truth.

Why is blood thicker than water? Literally, because of the various "formed elements" it contains: white blood cells, platelets, and especially, red blood cells. Figuratively, because it is full of genes.

One of the charms of anthropology is that the vast diversity of how people live provides an expansive view of what it means to be human. *Homo sapiens* is the world's most wide-ranging species. This is true geographically—people occupy just about every environment on Earth—and also behaviorally: human beings are as wide-ranging in *how* they live as in *where* they do so. They can be nomads, hunter-gatherers, pastoralists, farmers, or industrialists, high-tech software developers or low-tech ditch diggers, fierce warriors or nonviolent pacifists, monarchists, democrats, or anarchists. They adorn their bodies in all sorts of strange ways, worship every tree or a single omnipotent Sky-god, and they live in a kaleidoscopic variety of social systems and local customs. They speak a Babel of languages, make music and mathematics, and can live in space stations, submarines, tropical rainforests, the Antarctic ice continent, and the Saharan desert. People seem able to live anywhere, and in almost any way.

Notice, however, the word "almost."

Because the truth is that underneath the fascinating variety of different ways of being human there resides a substratum of similarity. It is as though a baker took the same chocolate sponge cake and then adorned it in with all sorts of fancy icings, applying tasty squiggles of almost infinite diversity. Underneath, however, there remains the same old stuff. No matter what flavor of cultural uniqueness is written on the icing, the cake says one thing only: HUMAN BEING.

Archaeologists make much of ancient relics—remnants of bone, shards of pottery, flakes of stone—to trace humanity's history. Another treasure trove of ancient *Homo sapiens*, often overlooked but far richer than the usual artifacts that fascinate archaeologists, is the human being of today. Some of these reliquaries are anatomical: vestigial structures such as our appendix or perhaps our tonsils. Some are physiological, embedded in the ebb and flow of chemical and electrical events. Most interesting for our purposes is behavior, stigmata of our animal past which even modern, high-tech human beings carry about as they rush toward the twenty-first century. And prominent in this evolutionary baggage is the human propensity for gene-based altruism.

There is a phrase for this genetic luggage: human nature. And not everyone believes in it. Some argue that human beings are entirely shaped by their experiences. If they are right then people are putty and there isn't much to being human. Denied human nature, no one really *is* anything, beyond an empty vessel waiting passively to be filled by whatever the environment has in store. Everyone would then be at the mercy of those movers and shakers and shapers who control the various "inputs" that make people what they are. "You are imagining that there is something called human nature which will be outraged by what we do and will turn against us," says O'Brien the Thought-Policeman, to Winston Smith, in George Orwell's nightmare vision of totalitarian control, *1984.* "But we create human nature. Men are infinitely malleable."

O'Brien is wrong. People are immensely malleable, more so, in all likelihood, than any other species. But *infinitely?* No way.

It is hard to avoid being dazzled by the diversity of ways in which human beings go about their lives, but it is also striking to recognize that many of these riveting details are in fact variations on a limited number of themes.

Attempting to make sense of gene-based altruistic behavior (not to mention the meaning of life!), this book shall proceed in a logical and—if successful—persuasive way, from theory, to animal examples, to cross-cultural consistency among human societies that are otherwise very different, to the day-to-day lives of modern Americans. Chapter 1 reviewed the theoretical argument for a gene's-eye perspective on evolution and behavior. Chapter 2 narrowed that focus to altruism, and took a rapid tour through its manifestation in animals. The present chapter looks specifically at altruism in human beings, first through the cross-cultural lens of anthropologists, and then bringing that focus closer to home.

Nepotism—And Its Inverse—Around the Globe

You are a stranger, just arrived at a settlement of Australian aboriginals. You wait outside in the hot sun, more than a little nervous while an old man—well versed in genealogies—is summoned. He interrogates you; if you are found to be related to anyone in the camp, you pass the test, and are allowed to enter. If not, you are killed.

The Tiv people of Nigeria are ready to expand their territory. Which of their neighbors do they attack? The most vulnerable? Those who occupy the most desirable lands? No, they invariably fall upon the adjoining group to which they are most *distantly* related.

The Nuer people of the southern Sudan have no formal government. As recently as several decades ago, they were almost constantly at war with each other, except when they united to raid another tribe, the Dinka . . . who were not their relatives.

There is an old Arab saying: "Me against my brother; my brother and me against our cousins; me, my brother and my cousins against our nonrelatives; me, my brother, cousins and friends against our enemies in the village; all of the village against the next village."

One way to understand the evolutionary genetics of altruism is to look at altruism's ugly doppelganger, victimization. Here, the pattern is clear: given the choice, people would rather prey upon those who are *un*related to themselves, or if forced by circumstances to attack their kin, they single out the most distant relatives for the most violent consequences. But wait! How about the old refrain, "You always hurt the one you love"? Bear in mind, first, that "love" has many different meanings, one of which is a spouse or lover, with whom genes are notoriously *not* shared. And second, even when people do harm their genetic relatives, this must be seen in the context of the immensely greater amount of time—and often intense emotions—shared with these same relatives[1]. In his poem "Crazy Jane Talks with the Bishop," W. B. Yeats gestured toward the ambivalent complexity of human family bonds: "Fair and foul are new of kin/And fair needs foul . . . /For nothing can be sole or whole/That has not been rent."

It comes as no surprise that if there is any truly shared concern, found among all human beings on this planet, it is genealogy. Not the arcane tracing of ancestral heraldry but a deeply felt need to situate one's self within the fabric of humanity, to know who we are in relation to others. And this concern with "relationship" is not simply defined by social relationships, but rather, by patterns of biological— that is, genetic—connectedness. (Many social relationships are predicted by genetic relationships, and that is exactly the point!) Then, having established their genetic identity—more accurately, after defining their social identity in terms of their genetic identity—people proceed to favor their relatives.

In fact, it is virtually impossible to imagine a culture in which people did not organize their lives and their concerns around genetic connections. It is easy to picture living without television, or microwave ovens, or to design a society in which everyone worked at home (or away from home), or in which everyone slept on waterbeds, or on the floor. But it simply goes beyond human possibility to introduce utter disregard for biological relationships, in which, for example, strangers regularly intrude themselves into each other's personal living space, in which any randomly chosen individual was as likely as one's parents to pay for college tuition, in which people did not introduce a biological bias toward others.

One of the best-known demonstrations of this genetic favoritism comes from field observations by Napoleon Chagnon and Paul Bugos, anthropologists who were on the scene (along with a cinematographer), when a now-famous "ax fight" broke out in a Yanomamo village. The Yanomamo, it should be explained, are indigenous inhabitants of the Orinoco basin rainforest, in Venezuela and Brazil. They garden, hunt, and are renowned—by their own account—as "the fierce people." In this particular incident, a group of visiting Yanomamo had been overstaying their welcome. They had relatives within the village, however, and claimed that as a result, they were entitled to be fed and housed longer than tradition would otherwise dictate.

As tensions rose, one of the visitors encountered a woman from the host village, who was carrying plantains home to her family. The visitor—named Mehesiwa—demanded some of the plantains. The woman refused; Mehesiwa beat her. She ran screaming to the village, whereupon her half-brother, Uuwa, rushed at Mehesiwa, loudly insulting him and gesticulating with a large club. Mutual insults were hurled, weapons were brandished; supporters of each man lined up on one side or another. Outright hostilities soon erupted, with additional men becoming involved. Fortunately, only the blunt side of axes and machetes were used, but at least one young man was knocked unconscious, while the two groups glared ferociously at each other. The two sides, supporters of either Mehesiwa or Uuwa, eventually disengaged and shortly thereafter, the visitors went home.

By carefully studying the filmed record of this incident, Chagnon and Bugos—who knew the geneological relationships of all participants—were able to assess the impact of kinship on "who supported whom." The results were undeniable: On each side, relatives lined up against each other.[2] Earlier, Chagnon had also found, interestingly, that Yanomamo villages are especially likely to split into two or more sub-groups when the average genetic relatedness among village members becomes particularly low. Not only that, but after splitting, those who stay together tend to be more closely related and to enjoy greater amicability than had been true in the larger group of more distantly related individuals, a pattern almost identical to those discussed for macaque monkeys in chapter 2.[3]

Turning now from antagonism to amicability, one thing stands out as a human universal: nepotism. Everywhere, people favor their relatives. Not only that, but the closer the relative, the greater the favor. When an Inuit whaling chief places his cousin in the umiak most likely to make a kill, or a !Kung san ("bushman") hunter makes sure that his brothers get first dibs from the innards of a freshly killed wart hog, or the CEO of a large corporation finds a fast-track job for his nephew—all are acting in concert with a universal, altruistic (and genetically selfish) imperative, one that many observers may decry as "unfair" even as, deep inside, they understand perfectly.

Ever since William Hamilton opened their eyes to the evolutionary biology of altruism, a growing cadre of biologists and anthropologists (abetted, on occasion, by some sociologists and historians) have been having a field day, examining how people behave toward each other, and finding—time and again—that human beings do not dispense altruism randomly. Rather, people favor their relatives. Here are just a few more examples: A research paper titled "Support and conflict of kinsmen in Norse earldoms, Icelandic families and the English family," examined historical documents and Icelandic sagas. The authors concluded that "close biological kin were more likely to support one another and, when conflict did occur, were less prone to kill one another than were more distant biological relatives."[4] Another, "Kin networks and political leadership in a stateless society, the Toda of South India," found that leaders are individuals with "positions of centrality in a kinship network."[5] Here's one more, chosen almost at random from the virtual avalanche of kin-oriented research studies that have appeared in recent years. "Relatedness and mortality risk during a crisis year: Plymouth colony, 1620-1621," reported on a disastrous winter, when 52 percent of the new arrivals on the Mayflower succumbed to malnutrition and disease. Its conclusion? "[T]he presence of relatives in the colony played a major role in survival." Relatives were especially likely to share food, provide medical aid, and so on. Those without relatives were alone indeed. [6]

Take a leap, next, into the twentieth century, landing on the Maine coast, where lobstering has long been a way of life. It is also a fiercely competitive occupation, not uncommonly leading to violence when rival groups defend their harvesting territories. In this world of competing "lobster gangs," information is crucially important—especially, information about where the lobsters are especially abundant—and it is not shared readily. There is, however, one notable exception: "Very close relatives exchange accurate information regardless of skill level. Fathers, uncles, and grandfathers feel obligated to help sons, nephews, and grandsons."[7]

A study of the daily lives of 300 women living in Los Angeles, aged thirty-five to forty-five, examined their experience of giving and receiving aid. What sort of aid?

When I needed money to get into the union; When I broke my collarbone and he took over the house; Talking to a friend about her marital problems; Picking up a friend's kids the whole time she was sick; When my son was in trouble with the police; She kept the children when my third child was born; When her husband left her; When she had a leg amputated; Loaned us money for a house down-payment . . .

The women were interviewed about the approximately 2,500 times such assistance was given and received. The findings were consistent: the closer the genetic relationship, the more altruism.[8]

In a different research project, respondents in both the United States and Japan were surveyed about how they would likely respond if they could only save one of three people in a burning house. The imagined victims, who were also potential recipients of altruism, included full siblings (genetically related by a factor of 1/2), and others, to whom the subjects were related by factors of 1/4 and 1/8. Once again, the closer the genetic relationship, the greater the reported inclination to help.[9]

Research of this sort must be seen as incomplete, especially since it focuses on surveys—what respondents say they would do—as opposed to what they actually do. As we shall soon see, however, people are in fact more likely to rescue relatives than non-relatives, at least as measured by recipients of the Carnegie Medal for heroism.

Just as alarm calling among animals can be interpreted in many ways, of which kin selection is just one (albeit an especially compelling one), the fact that people favor their relatives can also be interpreted in many ways. For example, maybe people are likely to be especially generous, even altruistic, toward their kin because they are likely to know their kin better than strangers. Furthermore, nepotism doesn't appear in a social vacuum, and in fact, cultural rules and family traditions typically urge people to treat their relatives with special care and kindness. It is possible to argue, therefore, that nepotism is undergirded by social factors after all, with biology a mere "epiphenomenon," like the froth on a wave, something that is just along for the ride.

There is another way to look at it, however, a way that includes both biological *and* traditional conceptions of cultural causation. It is simply this. When it comes to nepotism, people are almost certainly influenced by who they know, and by the dictates and urgings of social rules. Only a fool would deny this. But it is also fair to ask why people, around the world, are especially likely to know their relatives, and why social rules, world-wide, are especially likely to urge people to bias their behavior in favor of these relatives. In short, much of nepotism's immediate causation may well reside in our culture, whereas much of the reason why cultural traditions are what they are, appears to reside in our shared biology. It may simply be a matter of how deeply we care to dig.

For everyone, and not just lobstermen and Los Angelinos, family ties are of immense practical importance, not simply theoretical or sentimental. And not surprisingly, research has shown that when older Americans need help, they turn especially to blood relatives. Moreover, as the need for assistance increases, they are proportionately more likely to seek aid—and to receive it—from kin than from non-kin.[10] And to no one's great surprise, studies of grieving following death of a family member show that the closer the relative who has died, the greater the grief.[11]

Maybe these examples are overkill. The tendency of people to be nice to their relatives and to mourn their passing is something just about all of us take for granted, leading some, perhaps, to suppose that all this kin-biased behavior that evolutionary biologists have been so busily documenting is a tempest in a teapot: *"Of course people practice nepotism; that's human nature."* Precisely: that's the point. Thanks to the gene's-eye view provided by kin selection, we can finally begin to understand *why* nepotism is so important a part of human nature. In the same way that an intelligent fish almost certainly wouldn't describe it environment as "wet," nepotism is part of the water in which we all swim, something so obvious as to escape notice.

If you were to interview an altruistic, kin-oriented animal it probably would also deny that it is striving to make the most of its inclusive fitness, although most biologists have little doubt that animals go about doing just that. Rather, most nepotistic animals would say they do what they do because their close relatives are so cute, because it feels good to help them, because they just can't help acting in a certain fashion, and so forth. They might even point out that favoring their kin is a cherished tradition among their species, or in their family, something they learned in the nest or the den, at their mother's snout or their father's claw. If asked why they do something (marry, eat, sleep, fall in love, come in out of the rain, scratch when they itch) people, too, are unlikely to answer, "in order to maximize the promulgation of my genes." Similarly, if asked why they are favorably predisposed toward their relatives, they probably wouldn't respond like J. B. S. Haldane, performing a rapid mathematical calculation.

Yet, the calculation goes on, unconsciously but—it appears—unavoidably. Anthropologist William Jankowiak and sociologist Monique Diderich, both from the University of Nevada, Las Vegas, teamed up in a study of the inhabitants of a rigorously fundamentalist Mormon community known as Angel Park. Jankowiak and Diderich examined thirty-two polygamous families (in each of which one man had numerous wives), and in which strong efforts are made to obliterate any genetic distinctions among the children. In a polygamous household, all children are offspring of the father, and thus, all are half-siblings. However, only the offspring of the same woman are full-sibs, having not only the same father but also the same mother. Not surprisingly, it is in the father's interest to minimize conflict and maximize social harmony by inducing all the children to treat each other as siblings—full siblings, that is, and not to discriminate in favor of their mothers' children and against those born to the family's other women. But the researchers found "evidence for more solidarity between full siblings than between half siblings. Our data suggest that, despite the force of religious ideals, and notwithstanding the continued close physical proximity of half siblings in the polygamous family, there is a pronounced clustering of feeling and affection . . . that is consistent with inclusive fitness theory."[12] In short, individuals insist on preferring full sibs (50 percent genetic identity) to half sibs (25 percent).

Next stop: identical twins. They provide a unique and especially interesting situation. Genetically, they are 100 percent the same. We might therefore expect a remarkably high degree of altruism among identical twins. (On the other hand, since they are rare, the possibility exists that natural selection simply has not been able to work on such an unusual opportunity for altruistic self-promotion.) It is interesting to note that identical twins are not only remarkably similar in their behavior as well as appearance, they also seem to behave more cooperatively than do nonidentical—so-called "fraternal"—twins.

Maybe twins behave similarly because they are treated similarly; moreover, identical twins—because they look the same—are probably likely to be treated the same, or close to it. Accordingly, it is especially revealing to examine cases of identical twins who were reared apart (either through adoption or foster homes), and if possible, to compare them with fraternal twins reared apart. Fraternal twins are no closer, genetically, than any pair of siblings. They have, on average, one-half their genes in common; identical twins share all their genes. Fraternal twins reared apart are one-half as likely to share a variety of personality traits as are identical twins reared apart. And among these traits is empathy, close kin to altruism.[13]

Nancy Segal, a psychologist associated with the world-renowned

Minnesota Twin Study, compared identical and fraternal twins in their inclinations to work together while solving a simple puzzle. Her findings: Identical twins are significantly more cooperative and less competitive. She also noted that reunions of separated identical twins tend to be especially joyful, and she emphasized that whereas previous work has concentrated on the extraordinary similarities (physical as well as behavioral) between such twins, there remains a huge and unexplored avenue for future research, examining what appears to be the highly *altruistic* social relationship between them.[14]

Even marriage systems show what seems to be the imprint of kin selection. In the last chapter, we eavesdropped on the nepotistic love-life of wild turkeys and lions. Human beings are eerily similar. Thus, polyandry—the marriage of one woman to more than one man—is understandably rare, just as it is among animals. This is because a "reverse harem" doesn't especially benefit the wife (she can only have one child at a time, regardless of how many husbands she may accumulate), while it greatly diminishes the fitness of those husbands, each of whom would generally do better with his own wife instead of sharing one. But, as in the animal cases, if wife-sharing is to happen, then from the perspective of each husband's genes, far better if the men in question are closely related; in such a situation, as with turkeys and lions, a male can at least be an uncle, if not a father. Sure enough, in those rare cases of polyandry, among people as different as the mountain-dwelling Tre-ba of Tibet and the rainforest Nambikwara of Brazil and Paraguay, husbands sharing a wife are nearly always brothers, thereby sharing genes as well.

Whole armies of anthropologists have sought to identify and understand the kinship systems of the world's peoples. All because of a peculiar fixation—not on the part of the anthropologists, but of the cultures that they seek to understand. Kinship systems are a central concern of all human beings. Most of the social rules, from group membership, exchange and authority, to marriage and inheritance—in short, who does what to whom and with whom—are structured by kinship. Maybe this simply reflects a tendency to generalize from parent-offspring or sibling relationships to people-people interactions across the board. Or, as others maintain, maybe it's just a way for people to structure their social life. After all, some sort of distinctions are necessary: We can't just treat others like undifferentiated blobs of tapioca pudding. (As a matter of fact, we could: People don't *have* to be biased toward their kin, but being indifferent to relatedness would be to forego the evolutionary advantage of inclusive fitness.)

By and large, however, most anthropologists still view the human preoccupation with kinship as the arbitrary result of cultural rules and traditions, as though culture exists of and for and by itself, disconnected from the living creatures who create those rules and traditions. Incredibly, one respected senior anthropologist actually claimed that people couldn't possibly organize their social relationships according to genetic relationships because most human societies lack the symbols for fractions! (This is a bit like claiming that people can't think unless they know neurophysiology.)

Alternatively, there is a powerful explanation for why kinship is so important to human beings: because kinship is based on genes, and so is evolution. In their fine-tuning, the diversity of kinship systems do not always correspond exactly with what a trained geneticist might prescribe. Sometimes aunts, for example, are considered as genetically more distant than uncles. Sometimes it's the other way around. (In reality—that is, genetically—they are the same.) Sometimes, parallel cousins—the offspring of two brothers, or of two sisters—are distinguished from cross-cousins, the offspring of a brother and a sister (once again, parallel and cross cousins, although occasionally treated differently because of social traditions, are genetically equivalent). Sometimes first cousins and second cousins are considered the same (genetically they are not), and sometimes relatives by marriage are equated with relatives by descent. In the United States, there is frequently a cut-off at the level of first cousins; more distant relatives are typically lumped together as "distant cousins," as though a second cousin is the same as a first cousin, once removed.

But even though human kinship systems often violate biology in their fine details, to conclude that overall, kinship bucks biology would be to miss the forest for the trees, and in the process, to arrive at precisely the wrong conclusion. The basic, monumentally important fact is that people everywhere distinguish relatives from nonrelatives, and furthermore, they all make a distinction between close relatives and more distant ones. Not only that, but they all treat close relatives more altruistically than distant relatives. Sometimes, as we have seen, they make mistakes, just as people sometimes make mistakes in arithmetic; but everyone "knows" basic kinship, just as everyone knows that 2 + 2 = 4.

Exceptions that "Prove" the Rule

When it comes to evolution, on the other hand, some people insist that 2 + 2 = 5. For instance, Marshall Sahlins of the University of Michigan—a persistent opponent of evolutionary approaches to understanding human behavior—has pointed out that among many of the groups studied by anthropologists, people include a wide array of male relatives under the terms "father" and "brother," and likewise for "mother" and "sister." This seems to contradict evolutionary prediction, since a male first or second cousin, for example (distantly related via biology), should not be considered equivalent to a father or mother, which are biological near-kin. But in fact, when studies focus on what people *do* as opposed to the term of classification that they employ, it is revealed that nearly everyone knows perfectly well that his or her sisters, for example, are different—that is, "closer"—than cousins. Moreover, they are treated differently. From analysis of the renowned Yanomamo ax fight, for example, many anthropologists (although probably very few nonspecialists) were surprised to find that when push came to shove, villagers discriminated between a "classificatory brother" and a real—that is, a genetic—brother.

Americans are often exhorted that we are all brothers (or sisters) under the skin, just as citizens of the same country are often reminded that they are "children of the same fatherland," or "motherland," or of the same (generally paternal) god. But this doesn't mean most people don't also know who are their "real" brothers, sisters, or parents.

Another of those cases in which anthropology has been reported to "disprove" evolutionary genetics concerns so-called "mother's brother" societies. In these situations, fathers are notoriously uninvolved in the care of their children. Food and attention are lavished, instead, on the offspring of their sisters. (Hence the designation "mother's brother," because the primary male caretaker of children is not their father but their mother's brother.) These societies—and by some counts they constitute nearly 1/3 of all human social groupings—are regularly trotted out as seemingly incontrovertible disproof of evolutionary pressures, since it seems profoundly counter-(bio)logical that a man should prefer his nieces and nephews to his own children, when he is more closely related to the latter.

Upon closer inspection, this turns out to be an example of the exception proving the rule. Entomologist-turned-human sociobiologist

Richard D. Alexander, of the University of Michigan, gave the clearest evolutionary explanation of this seeming exception when he showed that it really isn't an exception at all, but is perfectly consistent with the algebra of inclusive fitness.[15] Alexander pointed out that if extramarital sex is common, then a man may well be unrelated to "his" (that is, his wife's) children. By contrast, such a man is guaranteed to share genes with the offspring of his sister: Her children are undeniably hers, and he and his sister are undeniably related through *their* mother. Sure enough, analysis of societies in which the mother's brother acts as the "father" reveal that they are likely to be those in which confidence of paternity is unusually low.

In an even earlier suggestion to this effect—reminiscent of Haldane's early appreciation of kin selection—anthropologist M. C. Kahn glimpsed the facts but apparently without fully understanding their significance:

> The head of the family is not the father, but the mother's oldest male relative—her brother or her mother's brother. The father is technically unrelated to his children. As a matter of biological fact, he can never be sure of his paternity, whereas maternity is never in doubt. All children belong to their mother's family, and are subject to the authority of her brother or her maternal uncle.[16]

The key concept is that extramarital sex makes a husband less likely to be the father of his wife's children. Combine this with expected gene-centeredness, and we get the following sociobiological prediction: "mother's brothers" societies are especially likely to be those in which extramarital sex is particularly frequent. In this way, men wind up lavishing parental care on children to whom they are at least minimally related. Here are some accounts of various "mother's brother" societies: Among the people of Truk, "extramarital affairs are practically universal." In the Dobu, "fidelity is very, very rare. Typically his wife will commit adultery with a village 'brother,'" he with a village 'sister.'" As for the Nayar people of India, "It is not certain how many husbands a woman might have at one time; various writers of the fifteenth to eighteenth centuries mention between three and twelve," not counting "occasional fleeting visits," so that "no Nayar knows his father."[17] Perhaps the most famous case of mother's brother affiliation was reported by the renowned anthropologist Bronislaw Malinowski. Among inhabitants of the Marquesa Islands—immortalized in Malinowski's now-classic *Argonauts of the Western Pacific*—men frequently went on lengthy oceanic trading excursions. The Marquesans,

interestingly, claimed that sexual intercourse was unconnected to pregnancy, and as proof of this contention were known to point trium-
phantly to the fact that a married woman not uncommonly produced a
child while her husband was at sea for several years!

Along these lines, a revealing study on inheritance in non-Western
societies was conducted by two biologically oriented anthropologists,
Steven Gaulin and Alice Schlegel of the University of Pittsburgh.
Gaulin and Schlegel divided various societies into two groups: those
where extramarital sex was extremely rare and in which men accordingly had a high confidence of paternity, and those where extramarital
sex was more common and men had a lower confidence of paternity.
The researchers found that in the former, fathers are much more likely
to bequeath resources to their wife's children, whom they assume to
be their own. In the latter, men are more likely to support their nieces
and nephews.[18]

In summary, it is striking that even though people often do not
know the basic facts of genetics, or sometimes even of human reproduction, they nonetheless act as though they do, not the least by developing family systems in which men invest in the same direction as
their evolutionary interest. Most people know relatively little about the
details of digestion, yet they get hungry and respond appropriately:
they eat. They may not know very much about respiratory physiology,
yet once again they do the right thing, several times per minute: they
breath. And whereas they don't necessarily know anything about evolutionary genetics—and may even vigorously deny its relevance—
they nonetheless behave as though they hear it whispering within them.

Some decades ago, a New York-based bakery used to advertise its
product showing Irish policemen, Italian grocers, and African American basketball players enthusiastically eating a rye-bread sandwich
under the caption "You don't have to be Jewish to love Levy's."
Similarly, you don't have to be an evolutionary biologist to act in
accord with your genetic interest.

Inheritance

An interesting way to test whether genetic "knowledge" is reflected
in peoples' actual behavior is to examine the details of inheritance. In
his dark tale of inherited guilt, *The House of the Seven Gables*, Nathaniel

Hawthorne referred to the tendency of people to pass their property in the same direction as their genes.

> There is no one thing which men so rarely do, whatever the provocation or inducement, as to bequeath patrimonial property away from their own blood. They may love other individuals far better than their relatives; they may even cherish dislike or positive hatred toward the latter, but yet, in view of death, the strong prejudice of propinquity revives and impels the testator to send down his estate in the line marked out by custom so immemorial that it looks like nature.

A biological updating of Hawthorne would suggest that the "strong prejudice" operating in such cases is not based on "propinquity," but rather, on genetics. (Or, that propinquity and shared genes are intimately connected, so that being influenced by the former, we also serve the latter.)

A century and a half after Hawthorne, a study of 1,000 probated wills in Vancouver, British Columbia, strongly confirmed that the custom of leaving property to one's "blood relatives" does not so much "look like nature," as it *follows* nature.[19] Inheritance patterns, it turns out, are strikingly in tune with kin selection theory; in particular, relatives are consistently favored over non-relatives or impersonal organizations, and closely related kin are favored over those more distantly related. To be sure, when making out their wills, people sometimes "disinherit" close relatives in favor of more distant ones, or even non-relatives. But this is rare, and because of its rarity—as well as, presumably, the fact that it violates basic "common sense"—it often receives quite a bit of attention when it occurs. Not surprisingly, whenever a wealthy old person leaves the bulk of his or her estate to an unrelated butler, or a favorite pet, this often leads to the will being contested. And by whom? Relatives. Moreover, any favoring of non-relatives over close kin is itself typically presented as prima facie evidence of mental incompetence!

But wait. What about this argument? *Of course* people generally leave their money to relatives, and indeed, to close relatives whenever possible: they are simply following the dictates of social tradition. The point is that social traditions are not random. Time and again, they point in the same direction as biology. As the wise rabbi said, it is possible for many different people to be correct.

Another interesting prediction supported by the Canadian study is that wealthier people leave money preferentially to sons, whereas those who are less wealthy favor their daughters. Since parents are equally related to their sons and their daughters, this predicted favoritism requires an explanation. Here it is. As first pointed out by Harvard University's Robert Trivers and Daniel Willard, males vary more than females in their reproductive success: Very successful males can and often do have more than one mate, whereas even the most successful females are comparatively limited, biologically, in the number of offspring they can produce. This leads to the prediction (confirmed in most but not all cases) that high ranking, dominant animals should be more likely to produce sons, whereas low ranking, subordinate animals are prone to producing daughters.[20]

When it comes to human beings, a similar pattern applies. A successful man can often turn wealth into reproductive success; by contrast, financial or social success among women is not as readily converted into evolutionary returns. Men are able to transform money into wives, or concubines, and thus, into fitness, via offspring, whereas women are less able to do the same with regard to husbands or lovers. After all, the biological limits on a woman's reproduction are far more restrictive than those operating on a man. Assuming, once again, that parents are interested in producing the greatest possible number of grandchildren—whether they consciously admit it or even know it—this leads to an evolutionary prediction: the greater the degree of polygyny in a given social group, the more inheritance of wealth should favor sons, whereas with less polygyny (and therefore, less opportunity of turning money into wives, and thus, children), there should be less inheritance-bias for sons over daughters. This prediction, too, has been confirmed, looking at a large cross-cultural sample of 411 societies.[21]

As the study of probated wills has shown, it appears that a similar pattern holds even within monogamous societies, such as Canada. Although social conditions have changed, the evolutionary "sweet tooth" of human beings may well predispose us to do things that served our fitness in the past. In this case: give wealth especially to sons if you are rich, to daughters if you are poor.

Recognizing Relatives

Let's grant that there is an evolutionary payoff for genes to help copies of themselves in other bodies, whether children or other relatives. But how do they identify such bodies? For biologists, this question has given rise to a new area of research known as "kin recognition." Not surprisingly, there have been many different answers. In some cases, animals focus on some aspect of themselves and then apparently try to match this trait with information derived from other individuals. The idea—sometimes known rather indelicately as "armpit sniffing"—is that a creature may smell its armpits, or listen to its own voice, or notice its particular coloration, and then behave with special benevolence toward other animals that most closely resemble itself.

Other possibilities include simply behaving altruistically toward neighbors versus strangers: all other things being equal, neighbors are likely to be more closely related. Or, a young animal might learn its relatives by observing the behavior of its closest relations: be nice, for example, to anyone that Mom treats nicely, or—in the simplest case—to anyone hatching in the same nest as yourself, or from the same litter.[22]

Finally comes an especially perplexing possibility, the capacity for genetic self-recognition: the ability to know, instinctively, who is your relative, without learning it, and without even sampling yourself and comparing stimuli. Many species of tadpoles have this ability, although to date, biologists don't really know what good it does them, or their genes.

An interesting variation on this seems to take place among beavers. They build small mounds of mud and leaves, which they anoint with secretions from their scent glands. Other beavers, passing through, may add their personal, odoriferous calling card (not unlike male dogs at fire hydrants). Biologists from the State University of New York College of Forestry at Syracuse looked at the response of beavers to pairs of such mounds, one of which was made redolent with secretion from a sibling, and the other, from a non-relative. The test animals behaved more territorially toward the odor of the non-relative; in short, it "upset" them more. Especially intriguing was this finding: The *mates* of test subjects—who weren't familiar with the scent donors or related to any of them—also responded less strongly to scent from the mates' brothers and sisters than to non-relatives. So, at least in the case of beavers, the recognition of relatives versus non-relatives must be learnable, and not only that, individuals learn to extend their tolerance to include their in-laws! It would be as though a wife, for example, was not only more tolerant toward her own relatives, because of knowing and "liking" their odor, but she could also identify her husband's relatives, by their smell, too—presumably because it reminded her of her husband—which also led her to be more tolerant of them.[23]

When it comes to human beings, it seems obvious that we learn our relatives by absorbing the social information of our family and local group. There may even be something here for the psychoanalysts. Given that people are predisposed to dispense altruism according to genetic relatedness—that, in short, it "tastes sweet" to favor one's close kin—there should be psychological mechanisms permitting and encouraging such behavior. The psychological process of "identification," could thus be based, biologically, on the benefits of appropriately directed altruism. According to Freud, "Identification is . . . the earliest expression of an emotional tie with another person."[24] The tendency to identify with another person involves seeing one's self in that other, something that makes biological sense as well. In fact, it *only* makes sense biologically. "You" do not exist in another; there is some probability, on the other hand, that copies of your genes do. And the closer the genetic tie, the higher is that probability. Parents, not surprisingly, are especially prone to identify with their children (and vice versa). As Sigmund's daughter Anna wrote, "We know that parents sometimes delegate to their children their projects for their own lives, in a manner at once altruistic and egoistic."[25] Thanks to revolutionary biology, we may also, at last, know why.

Altruism as Selfishness and Vice Versa?

When altruism shows up in situations outside of kinship, there is always the possibility that it isn't really altruism at all, but actually selfish behavior, as with the alarm-calling prairie dog who might—at least in theory—be manipulating the behavior of its audience for its own personal benefit. Among human beings, there could be a comparable selfish benefit: in this case, a payoff in being perceived as an altruist, even if one is actually selfish at heart. (Indeed, the more selfish someone is, the more inclined he or she might be to erect an altruistic facade.)

Altruistic favoritism toward relatives generally has a bad name. Anti-nepotism laws, for example, are almost as widespread as nepotism itself. On the other hand, plain unvarnished selfishness is even less laudable. Far better to claim that one is acting, self*lessly*, for someone else, thereby cloaking one's own greed or search for private benefit in the acceptable cloak of helping another: "This hurts me more than it does you," or, as Don Quixote's avaricious relatives sing in *Man of La Mancha*, "I'm only thinking of him."

In addition, there is always the appeal to group-benefit, which goes a long way toward diminishing the resentment otherwise associated with selfish gain. Powerful politicians, for example, are regularly lauded for years or decades of "public service," usually obscuring the fact that such self-sacrifice is typically associated with the immense personal satisfaction of wielding power over others, combined not surprisingly, with generous financial, social, medical and retirement benefits. (And not coincidentally, no small amount of other payoffs as well, at least for unscrupulous male politicians eager to take advantage of the abundant sexual opportunities that often come along with situations of power and prestige.)

There may be other factors leading to professed altruism, including, perhaps, its usefulness as a courtship display: by showing how altruistic and self-sacrificing one is, and thus, what a good parent and mate one would be, individuals might make themselves more attractive and a better catch.[26] Charles Darwin had been troubled by the existence of heroic self-sacrifice, such as a soldier dying to save his comrades. Specifically, he could not understand how the "offspring of the more sympathetic and benevolent parents, or of those which were the most faithful to their comrades, would be reared in greater number than the children of selfish and treacherous parents," as well as the seemingly evident fact that the hero "would often leave no offspring to inherit his noble nature."[27] Darwin was onto something, and not for the first time! However, he lived a century before William Hamilton and thus, he did not realize that acts of heroism could be readily explained if they were typically associated with a kin-selected return. Ronald Johnson of the University of Hawaii analyzed 676 acts of heroism which resulted in Carnegie medals between 1989 and 1995. He found that of 676 people rescued, 46 were biological relatives (7 percent); but in cases of severe self-sacrifice, when the rescuer died in the attempt, biological relatives made up fully 32 percent.[28]

Kin selection wasn't the only possibility that eluded Darwin. He never considered that perhaps heroism (or altruism) might be helpful in attracting mates. Here, another of Johnson's findings becomes relevant: 92 percent of Carnegie medalists were male. Perhaps men are inherently more courageous than women. More likely, men are socialized to see themselves as potentially brave and daring, and moreover, there is a likely biological wisdom in this teaching, and also, in all probability, in male susceptibility to it. Thus, men have long been rewarded with biological—that is, reproductive—success, as a result of successful displays of derring-do. (Recall that men typically have more opportunity than women to transform social success into evolutionary payoff.)

War heroes have historically enjoyed sexual rewards, thereby receiving a likely reproductive recompense as well. It is but a small step from being courageous to being altruistically self-sacrificing, especially since many people who end up being suicidally altruistic may actually have been attempting to be courageously altruistic, hoping to survive their altruism and come home to reap its benefits. If so, then much of the altruism that is so revered—and rewarded with posthumous medals and glory—may actually represent failed attempts at selfish achievement via altruism mixed with derring-do. Perhaps the secret wish of the Good Samaritan is that his or her altruistic act would be discovered and publicly acknowledged, but without the self-advertisement that might suggest the altruism is only being done for show or to attract mates, and thus may not be "genuine."

The same could also apply on a milder scale to other altruistic actions, such as generosity with money, with one's time and effort, and so forth.

But if altruism is often motivated, at least in part, by hoped-for success in obtaining a mate (or matings), why doesn't it disappear *after* marriage? To some extent, it does, as suggested by the worldwide preference for military inductees who are young and single. And what about youthful idealism? Could the biology of altruism help explain why people are often more inclined toward socially altruistic causes when young, and more selfish (that is, more conservative) as they age?

In addition, maybe a degree of altruism also helps in *keeping* mates, as well as in promoting the likelihood of an additional successful liaison on the side. Or, like the fondness for sweets, it may even be sufficient that people find it rewarding to be cheered and admired, even if the hero, altruist, or philanthropist is a confirmed bachelor, even a celibate priest. Bear in mind that as with most considerations of altruism, it isn't necessary for genetic goals to be conscious, or even for them to be often achieved in today's society; it would be sufficient if the behavior, and the mind-set that produced it, were effective through much of the human evolutionary past.

A Chemical Mechanism?

Bodies behave so that genes can get their way. Such is the view from revolutionary biology. Toward this end, a variety of immediate mechanisms may be employed, ranging from reflexes to complex social learning. When it comes to the tendency to be altruistic or selfish, an important intervening factor may well be "personality," which is simply another way of saying that people have predispositions to behave in one way or another. Dr. Robert Cloninger, a psychiatrist at Washington University in St. Louis, has suggested that there exists a spectrum of individual variation in "reward dependence," or need for social approval, mediated by a single chemical neurotransmitter, norepinepherine.[29] Highly "reward dependent" individuals are described as "industrious, ambitious overachievers who push self to exhaustion; highly dependent on emotional supports and intimacy with others; and extremely sensitive to rejection." These are the classic codependents, who seem to lack appropriate self-regard and self-protective mechanisms, and who sacrifice themselves on the altars of other people's troubles and inadequacies.

Coupling such characteristics with another personality type—termed "high novelty seeking" and mediated by another neurotransmitter, dopamine—Dr. Cloninger describes a "gullible-heroic" personality, who may engage in "excessive heroic risk-taking, such as risking own life to save total strangers."

In other words, Cloninger's theory may suggest a proximal mechanism, a testable, biochemical clockwork model, through which the genetics of altruism could operate. Certain individuals who are high in norepinepherine may be "people pleasers" to the core, so dependent on signals of social approval that they overspend their personal resource budget in the service of others. Such people, with their high degree of reward dependence, may achieve increased inclusive fitness much of the time. For example, they may be devoted parents, supportive siblings, generous and helpful cousins, and so on. But by caring so much for the affairs of others, they may also shortchange themselves when times are rough, or when others pick up on the opportunity to profit at their expense.

By contrast, an individual low in reward dependence, and hence cynical, solitary, and tough-minded, may be more indifferent to children or other relatives, yet under certain circumstances, more reproductively successful than the people-pleaser.

Cloninger's theory—which, incidentally, he does not explicitly link to evolutionary genetics—could also serve as an excellent example of how a trait may have a wide range of expression, with a corresponding range of adaptive advantages. It may yet prove that a cocktail of psychoactive chemicals, whose recipe is encoded by a limited number of specific and identifiable genes ("36 nannograms of dopamine, 8 nannograms of serotonin, shaken but not stirred . . . ") is responsible for the spectrum of altruism, from Scrooge to Santa Claus, from draft-dodgers to winners of the Congressional Medal of Honor.

Incest

If people are inclined to behave altruistically toward relatives, especially close relatives, then here's a question: Why don't they mate with their own nearest relations—the closer the better—thereby producing offspring who are genetically as similar as possible to themselves? Wouldn't this be a good way for genes to promote their own success? If a brother and sister have children together, for example, then each parent would not only enjoy the 50 percent genetic similarity of a "normal" mother or father to its offspring, but would get the extra evolutionary benefit of being an uncle or aunt as well (via the additional genetic connection achieved through his or her sexual partner). The same mathematics ought to apply to father/daughter or mother/son unions.

It seems that breeding among close relatives should be the norm. And yet, incest is exceedingly rare, and when it occurs, nearly everywhere it is viewed with a degree of horror; the so-called incest taboo is in fact another cross-cultural universal, on a par with nepotism.

For a time, the prohibition against incest was seen as uniquely human. Animals, it was widely held, were not so fussy. Only humans were granted the self-restraint and moral judgment to refrain from making love with those they love the most. In doing so, they supposedly established the foundation for civilization, and for our very humanity. Anthropologists and sociologists elaborated the social advantages derived from incest avoidance, notably the opportunity to establish alliances with other family groups. Psychoanalysts, beginning with Freud, erected an entire conceptual scheme on what they saw as the uniquely human need to overcome and redirect Oedipal tendencies as part of healthy development. The only problem with such formulations is that they didn't reckon on biological reality.

There are three major aspects to this reality: first, the simple fact—abundantly demonstrated in recent decades—that human beings are *not* unique in avoiding incest. Nearly all animals similarly refrain. Often, young individuals avoid mating with their parents or siblings by dispersing (leaving the area where they were born or hatched) before they become sexually mature. When, for some reason, dispersal is inhibited, play patterns will frequently interfere with sexual behavior among sibs, and in a number of species, sexually mature females will simply not ovulate when in the presence of their father. If anything, human beings may be unique in that, occasionally, they *do* practice incest!

Second, there is a straightforward mechanism that accounts for why incest is typically so rare among human beings: negative sexual imprinting. In brief, people who grow up together, beginning in early childhood, generally are loathe to view each other as potential sexual partners (no matter what the rules of society). Familiarity does not necessarily breed contempt, but neither does it breed a tendency to breed. The clearest demonstration of this process was by the late Israeli sociologist, Joseph Shepher, who studied marriage patterns among kibbutz residents. Here, children who had been reared together since toddlerhood showed a pronounced disinclination for sexual relations as adults. In fact, Shepher found that out of 2,769 marriages between second-generation kibbutz dwellers, only six took place within the peer group of any one kibbutz, and this despite the fact that such marriages were actively encouraged by authorities. When asked about possible sexual urges toward their peers, many young (and otherwise sexually active) kibbutzim explained that they couldn't take so-and-so seriously as a lover, because they were much more inclined to see him/her as a brother or sister. One responded that after years of growing up together and sharing the same potty with someone, it is almost impossible to think of her with sexual interest![30]

Another revealing case comes from the form of traditional Chinese marriage known as *shim-pua*. Here, the social norm for wealthy Chinese families was to adopt prospective daughters-in-law while they were still young girls and to raise them alongside their future husbands. Research by Stanford University anthropologist Arthur Wolf found that this arrangement was not very successful: In nineteen such adoptions that took place in one Taiwanese village, seventeen of the would-be happy couples refused to consummate their marriages, despite the urging of family and friends. In the two successful marriages, interestingly, the girls were older than usual when adopted (almost adolescent rather than young children) and so apparently did not develop pseudo-sibling relationships with their future husbands.[31]

Actually, it was a self-taught Finnish anthropologist named Edward Westermarck, in a 1891 book titled *History of Human Marriage*, who first proposed to the scientific community that being reared together leads to sexual aversion, which in turn provides a mechanism for incest avoidance. But hints of the "Westermarck effect" go back even earlier. In Jane Austen's 1814 novel *Mansfield Park*, for example, the argument is made that one sure way to prevent a rich man's sons from falling in love with their impoverished cousin is for the young girl to live with them, beginning as a child:

> breed her up with them from this time, and suppose her even to have the beauty of an angel, and she will never be more to either than a sister.[32]

So, we have the biological ubiquity of incest avoidance as well as a likely immediate mechanism for achieving such avoidance. The only thing remaining is the adaptive significance, and that is readily provided by genetics itself. The evidence is now overwhelming that breeding among close relatives produces offspring with a higher than average probability of genetic defects. This so-called "inbreeding depression" takes place because harmful genes, present in single dose, are more likely to encounter identical copies of themselves when close relatives mate. When this happens, the fitness-lowering traits—hidden when solitary—are liable to be expressed when there is no healthier, dominant alternative to cover up their effect. The upshot of all this is that incest is, evolutionarily, a bad idea. Not surprisingly, cultural traditions have strongly discouraged such behavior, and as we have just seen, it appears that the genes themselves have developed a generally effective mechanism to prevent it as well: sexual disinterest among people who grew up together from early childhood. In most cases, such people are likely to be siblings (the Israeli kibbutz and Chinese *shim-pua* are cases in which cultural tradition essentially fooled Mother Nature).

For the sake of completeness, it must also be confessed that the earlier discussion of incest was actually a bit misleading. Incest is unlikely to be favored by kin selection in any event, even though it might seem superficially likely. To be sure, an individual who mates with his daughter will produce offspring who have 75 percent of his genes. But these genes would also be projected into the future if the would-be incestuous parent and his daughter were each to reproduce, non-incestuously. Moreover, in such a case there would be no fitness reduction because of inbreeding. There may be a slight convenience in reproducing with a close relative and thereby concentrating one's genetic interest in a small number of individuals, but mathematically, no ultimate advantage. Add inbreeding depression, and the equation becomes distinctly negative.

So, any way you slice it, incest is a biological liability, not an opportunity for fitness enhancement via kin selection. And not surprisingly, it has generated responses that appear to be every bit as biological as they are cultural or societal.

Squabbles over Siblings, Stepsibs, and In-Laws

"Am I my brother's keeper?" Although God never directly answered Cain's question, it seems clear that Cain was certainly not supposed to be his brother's murderer.

In fact, siblings generally *are* expected to be each other's keeper. They are supposed to care for each other, help each other, cooperate and if need be, sacrifice for each other; in short, to display a whole lot of altruism. There are exceptions to this generalization (in chapter 6, we'll examine some of them). But it is when siblings do *not* behave altruistically—and especially when they act downright nasty toward each other—that tongues wag and people shake their heads in surprise. When Unabomber Ted Kaczynski was arrested, media attention focused not simply on the arrest of a major criminal, but on the riveting fact that he had been turned in by his brother, David. Sibling rivalry is real, and—as we shall see—understandable in evolutionary terms. Even more real, however, is the simple fact that siblings are expected to be protective of each other, and that moreover, they usually are.

It is probably noteworthy that when (and if) sisters or brothers become more distant, it is often in the context of their own reproductive activities. Having become adult, with their own spouses and children, their own workaday lives, siblings commonly drift apart, at least somewhat. The push of being a father or mother, husband or wife, can outweigh the pull of sibling bonds, but it is probably significant that it takes so deep and primal a gravitational force to rival that of brotherly or sisterly love.

By contrast, in hunter-gatherer bands and among many pastoralists, siblings often remain closely associated, frequently for their entire lives. One of the most powerful and generally unacknowledged stresses of modern Western life may reside in the conflict between being a "successful" reproducing adult and being a loving and attentive brother/sister (or son/daughter for that matter), especially when adulthood requires that siblings and other close relatives live geographically separated from each other.

At the same time, even as extended families are becoming increasingly rare, it is not uncommon for a modern wedding or graduation to include—in addition to the bride and groom (or the graduate) and parents—an array of stepmothers and stepfathers, often with a heavy load of animosity between the various divorced members of the older generation. In such cases, families are not so much extended as distended, subjected to peculiar and confusing pressures, part biological and part cultural. Since they are family members—and thus, kin of a sort—ex-spouses, ex-mothers-in-law, etc., are acknowledged part of the immediate social system, if for no other reason than because they remain important to their biological offspring. But for those to whom they are not related (or were "related" only through a marriage that is now defunct), the connection is much more distant, if not overtly hostile.

Such pressures are not restricted to the consequences of divorce. Even happy marriages are, in a sense, biologically tense, if only because they involve the joining of two individuals who in most cases are *not* genetically related. Even if the married couple has a genuine interest in each other—both through love and, not coincidentally, through their shared reproductive interest—the same cannot be said for the relatives of the husband and wife. (Actually, a shared reproductive interest still exists, but is at least one step removed from that of the married couple.) Not surprisingly, relations with in-laws are traditionally strained, not just in the United States but around the world. Mother-in-law jokes, for example, are well established in the lore of cultures as diverse as those of East Africa, New Guinea, and the Inuit ("Eskimos").

In-laws are usually identified as relatives, but of a special sort. Even when the distinction is not overt, it is nonetheless there. Modern readers, for example, are often confused by some of the nineteenth-century novels of Jane Austen, Charles Dickens or George Eliot, in which someone whom we would call a brother-in-law is referred to as a "brother," and a sister-in-law as a "sister." But as with the traditional societies studied by anthropologists, once we look beneath the surface, it is obvious that the characters in question know precisely who is related to whom, and in what way. When a Jane Austen heroine gushes deceptively, at the prospect of a marriage between her brother and a woman who is that heroine's chief female rival, "How lovely it will be to have you become my sister," there is no doubt that our heroine reserves a clear distinction between the prospect of a "sister" via marriage and one via "blood."

In this era of AIDS and other such diseases, it is sobering to acknowledge that sex with someone involves, in a microbial sense, intimacy with everyone else that person has ever slept with. Similarly—and perhaps marginally less frightening—is the fact that to marry is to establish a relationship not just with one's new wife or husband, but with your spouse's entire family as well. They become relatives of a sort . . . yet not quite.

In one way or another, people all over the planet "get married." The result is that a whole new constellation of characters enters into their lives, and these are genetic strangers, to boot. Kinship relationships are complicated enough, without this wrinkle. In an effort to simplify, perhaps, many people often refer to their relatives by marriage with the same terms that they use in referring to their "real," genetic kin. It is also possible that by tapping into a well of benevolence and comprehension associated with genetic kin, they are seeking to smooth over what is inevitably a difficult addition to their daily lives.

Anthropologists call it "fictive kinship," in which people apply the words and imagery of genetic relationship in order to evoke responses to non-kin which in some ways parallel those that arise more "naturally" toward genetic relatives.

These patterns of make-believe kinship are often used in other ways, too. Look for pseudo-gene terminology whenever the intent is to generate feelings of solidarity and altruism, especially when such responses aren't otherwise forthcoming. "Brothers" and "sisters" are called upon to stand up for one another, particularly when the individuals in question aren't really siblings at all. (If they were, such exhortations wouldn't usually be needed!) Not that the result is always pernicious; there is much to be said for universal "brotherhood," never mind how it is achieved, and even if we have to fool ourselves a bit in order to grease the psychological skids with a dollop of biologically potent rhetoric.

In other cases, however, the human penchant for kin-directed altruism leaves people vulnerable to dangerous manipulation. In country after country, citizens are exhorted to support the Fatherland, defend the Motherland, to rally round their brothers and sisters in Bosnia, Serbia, Croatia, Rwanda, Cambodia, Russia, indeed, anywhere people and their susceptibility to fictive kinship can be found. Which is pretty much everywhere.

All this supporting, defending, exhorting and rallying can be seen as altruism of a sort, even if it typically springs more from a biologically generated Achilles heel than from an accurate, clear-eyed assessment of genuine inclusive fitness interests. At the same time, it requires that we turn to this dark side of altruism: aggression, especially toward non-relatives.

Friends and Foes

What about violence *within* a group, even within a family? And what of that most extreme form of violence, murder? It goes virtually without saying that even when aggressive or downright violent behavior may be applauded toward non-group members, different rules and expectations apply within the group. Nonetheless, murderous violence is not reserved only for members of other groups, or even for non-relatives. Following a homicide, family members are prime suspects, and for good reason: Disproportionately, family members kill one another. ("Cherchez la femme" is often less useful than "cherchez la famille.") This certainly seems to fly in the face of the expected biology of altruism. And yet, when Canadian psychologists Martin Daly and Margo Wilson of McMaster University looked at homicide through a finely polished Darwinian lens, they discovered that far from being violated, the expectations of kin selection were supported to a remarkable degree.[33]

Daly and Wilson found, for example, that blood kin—so-called "consanguineal relatives"—accounted for only a small total of murder victims; most of the time, when relatives were killed, they were relatives by marriage, not by genes. And Daly and Wilson also recognized that when genetic relatives *do* kill each other, we must take into account their greater opportunity to do so, the straightforward likelihood that close relatives may rub each other the wrong way simply because they are so much more closely associated. (In the same vein, most automobile accidents take place within a few miles of people's homes, not because roads are inherently more dangerous near one's residence, but rather, because this is where most people do most of their driving.) This correction made, Daly and Wilson found that consanguineal relatives are in fact far *less* likely to kill or be killed than are non-genetic "relatives" who may be living in the same house. Moreover, they also discovered that when people get together and plan to kill someone, these coalitions are consistently liable to be composed of kin, conspiring to do violence against—you guessed it—non-kin.

Speaking of murder, here is some more food for thought: When insane people commit murder, they are more likely than sane people to kill close family members. To be insane is to be out of touch with reality. Part of reality—the biological part—is that a person's long-term evolutionary interests are closely interwoven with his or her genetic relatives. Significantly, the insane are out of touch with this aspect of reality; they don't know where their interests lie. (There is a danger of circular reasoning here: If people are diagnosed insane *because* they kill relatives, then this couldn't fairly be taken as evidence for the underlying point. However, there is strong reason to believe that people who kill their relatives really are unbalanced. For example, they are also more likely to commit suicide.)

It is also interesting that criminal penalties are typically *less* severe for people who murder their close relatives, perhaps because the judicial system tends to recognize that killing a kinsman is itself evidence of diminished mental competence, and/or because of the widespread feeling that in such cases, the perpetrator has "suffered enough."[34] The presupposition is so strong that people will not kill their close kin that when they do so—as with the Menendez brothers who killed their parents, or Susan Smith who killed her two sons—the media's response is unremitting fascination and a presupposition that there must be something especially wrong.

The history of murder is also revealing. Thus, only in recent times has it been considered a crime against the state. Among many different human societies even today, and for much of the history of Western Europe as well, murder was seen quite differently: as a crime against the victim's *family*. It usually required restitution more than punishment. This could be achieved either via retaliatory killing, or through some sort of direct payment, typically giving money or goods to the aggrieved: the victim's family. And significantly, the retribution or repayment was to be made by the *family* of the perpetrator. Even today, we speak of the criminal "paying for his crime," or "paying his debt to society," phrases that hearken to an earlier time, and that are redolent with no small whiff of evolutionary logic.

In-Group, Out-Group

Sociologists have a convenient epigram: "in-group amity, out-group enmity." Why are people benevolent toward group members? The answer is simple: in-group amity. And why are they often nasty and even violent toward outsiders? Out-group enmity.

Nearly 2,500 years ago, Aristotle had an equally convenient explanation for why objects fall: The ground is the natural place for all things, noted the great philosopher, and so, they seek it. Moreover, this also explains why things accelerate as they plummet to earth: Objects become increasingly "jubilant" as their unification with the ground grows more imminent, and so, they travel ever more rapidly. Thanks to Isaac Newton, "gravity" has replaced "jubilation," and although cynics might claim that one linguistic convention has merely superseded another, the truth is otherwise. After all, Newton's law of gravity enabled physicists to make detailed predictions, to test these predictions against observation, and to enter a whole new realm of understanding.

Speaking of new realms of understanding, kin selection combined with fictive kinship helps explain the powerful force of pro-social gravitation ("in-group amity") that acts on those special objects known as human beings. It also helps explain the contrary, repulsive force ("out-group enmity") that works the other way. The key is not jubilation, but genetics, the recognition that most natural groups are made up of individuals who share a higher-than-average degree of genetic relatedness among their members. At the same time, don't forget that the distinction between in-group and out-group was probably even more extreme in the prehistoric environments that constituted 99 percent of humanity's evolutionary history, during which people lived in small, relatively isolated social units, and within which genetic relatedness was undoubtedly much higher than it is in, say, a modern town, city, or nation.

In a very different world—of nuclear weapons, organ transplants, nation-states and the Internet—human beings nonetheless retain a powerful fealty to their group, combined with opposition toward other groups. Human beings are inclined to strive against each other, if only because each is genetically distinct. Each interest is separate, even as they come together in particular ways. And one of the most powerful of these comings together is the yearning for ethnic identity and association.

Bosnia, Somalia, Rwanda, Cambodia: these are not isolated instances of the power and danger of ethnic affiliation and conflict. Rather, they show us an important part of ourselves. Just as the most high-tech sports cars are still fueled by squashed dinosaur guts, it is a safe bet that third millennium, twenty-first-century human beings will continue to be fueled, emotionally, by allegiance to a primitive chant that repeats, deep in our bones, "in-group good, out-group bad."

Biological Commandments?

"Two things have always filled me with ever new and increasing admiration and awe," wrote philosopher Immanuel Kant. "The starry heavens above and the moral law within." Don't look to Darwin, even with modern updating, to explain the starry heavens above. But as to making sense of the moral law within, gene-focused evolution may go a surprisingly long way. In the process, even such apparently praiseworthy injunctions as the Ten Commandments may be revealed as less universal and less laudable than before. Take the fundamental Biblical injunction to altruism, "love thy neighbor as thyself" (Leviticus 19:18). With an evolutionary perspective, this statement, which among others so filled the great Kant with such awe, suddenly looks rather limited, parochial, and tribal. Here are three English translations and their dates:

Thou shalt not avenge, nor bear any grudge against the children of thy people, but thou shalt love thy neighbor as thyself. (King James Version, 1611)

> You shall not take vengeance or bear any grudge against the sons of your own people, but you shall love your neighbor as yourself. (Revised Standard Version, 1952)

> You shall not take vengeance or bear a grudge against your countrymen. Love your neighbor as yourself. (TANAKH, The Jewish Publication Society, 1985).

My point is that "neighbor" almost certainly means literally *neighbor*; that is, someone who is nearby, not just anybody and everybody, but an in-group member. Loving one's neighbor may turn out to be less a matter of true altruism than of genetic selfishness.

Turn to the beginning of the Ten Commandments (Deuteronomy 5:17-21, King James Version):

> Thou shalt not kill.

> Neither shalt thou commit adultery.

> Neither shalt thou steal.

> Neither shalt thou bear false witness against thy neighbor.

> Neither shalt thou covet thy neighbor's wife; and thou shalt not covet thy neighbor's house, his field, or his manservant, or his maidservant, nor his ox, nor his ass, nor anything that is your neighbor's.

As written, it seems that we are first enjoined from killing, then from adultery, then from stealing, each as a blanket prohibition, regardless of who is the victim. By commandment number four, we are told to refrain from bearing false witness against our *neighbor*, or from coveting our *neighbor's* wife, and so on. But as one scholar has pointed out, the original scrolls from which these words were translated contain no periods, commas, or capitalizations indicating the beginning of a sentence.[35] The King James Version, quoted above, is therefore very much a product of the translators' opinions and judgments. Given the flexible punctuation of the original, the Ten Commandments could just as well be organized as follows:

Thou shalt not kill, neither shalt thou commit adultery, neither shalt thou steal, neither shalt thou bear false witness against thy neighbor.

Read this carefully, for it is very different from the version quoted above, not only grammatically, but in its message. It is also more consistent with the actual wording, and very much in accord with the expectations of evolutionary altruism and kin selection. What it says is that we should not kill our neighbor, or commit adultery with our neighbor, or steal from our neighbor, and so on. By implication, then, its pretty much o.k. to kill a stranger, to commit adultery with a stranger, to steal from a stranger, and so forth. (Otherwise, why should we be more squeamish about bearing false witness against a neighbor than about, say, stealing from him or her?)

The point is this: Rather than a blanket prohibition against bad behavior across the board, the Ten Commandments can be interpreted equally—and probably more accurately—as a warning that such behavior is prohibited, *but only toward a fellow group member*. If so, then these things may well be acceptable if directed toward non-group members, who are also likely to be non-kin.

There is, in fact, no end of biblical urgings that the ancient Israelites kill and/or steal from all sorts of people, so long as the victims were strangers. Violent injunctions of this sort abound in the Old Testament, throughout Numbers (21:2-3, 21:34-35; 24:8; 24:19-20), Joshua (2:10; 6:21; 8:2; 8:24-26; 10:1; 10:28; 10:35; 10:37; 10:39-40) 11:11–14; 11:20-21), and especially Deuteronomy (2:34; 3:2-6; 3:21; 7:1–2; 7:16; 7:23-24; 9:3; 11:24-25; 20:16-17; 31:3-5; 33:27). And this is but a partial list. Here is a sample, from Deuteronomy 20 (Revised Standard Version)

> . . . you shall put all the males to the sword, but the women and the little ones, the cattle, and everything else in the city, all its spoils, you shall take as booty for yourselves; . . . But in the cities of these peoples that the Lord your God gives you for an inheritance you shall save alive nothing that breathes, but you shall utterly destroy them, the Hittites and the Amorites, the Canaanites and the Perizzites, the Hivites and the Jebusites . . .

The ancient Israelites were, by all accounts, a nomadic herding people, quite unrepentantly violent: toward out-group members, that is. In this regard, they were not very different from their neighbors, or from many other people today. Just as shared genes leads to Dr. Jekyll, a lack of such connectedness gives rise to Mr. Hyde. By digging into the anti-altruistic side of the Ten Commandments, I am not seeking to cast aspersions on religion in general or the Old Testament Israelites in particular, but rather to point out that all human beings are composed of part Jekyll, part Hyde, and that the Hyde-like "flip-side" of altruism—enmity toward non-relatives—is probably just as widespread as the benevolent evolutionary geneticist, Dr. Jekyll.

The Pyramid of Victimization

Nutrition experts speak of the food pyramid: complex carbohydrates at the base, fats and proteins toward the top. Here is a variation on this theme, a universal pyramid of victimization, by which all things are connected, genes not the least: At the base of the pyramid, human beings kill to eat. They unite against other species, cultivating or capturing them for food. Since no people (indeed, no animals) can make energy or protein from the sun, as plants do, everyone must eat—and therefore, kill—to survive and reproduce. Prey are called "food," and most people eat without guilt.

Moving along the food chain, however, as people confront species that are closer to themselves, with which they share more and more genes, the relationship changes. Exploitation gives rise to empathy, if not outright altruism. Even Buddhists recognize the need to kill and eat vegetables, while acknowledging that we do in fact share something ("Buddha-nature" in place of genes) with carrots and broccoli. However, we share more genes yet with other animals than with plants, and progressively even more genes with animals that are more like us, that is, more with cows than with shrimp, more with pigs than with fish.

Some cultures advocate complete vegetarianism. Others have complex dietary laws, forbidding certain plants or animals, with virtually no clear-cut bias *towards* evolutionary relatedness that we have been able to discern. Monkey meat is prized in Africa and parts of South America, and there are societies that still adore the flesh of gorillas or chimpanzees, when available. There are no worldwide prohibitions against eating primates, yet there are no cultures in which primates are farmed for food. There seems to be a widespread if not an "instinctive" preference for eating "lower" animals, that is, those to whom humans are more distantly related.

At least, there is less guilt. Interestingly, when "higher" animals—which means, essentially, those that are "closer" to humans—are eaten with special relish, such feasting often includes the cultural expectation that by consuming such meat, certain desirable traits will be transferred: courage from a lion's heart, sexual potency from a hippo's genitals, and so on. The clear presumption is that such conveyance is more likely when the creature being eaten is more similar to the consumer; once again, when there are more genes in common.

Cannibalism is no longer fashionable. Whereas it used to be relatively common for people to eat war captives, ritual sacrifice victims, or the deceased Big Man, these days people are more likely to eat Big Macs. However, it is interesting to note that cannibals were especially likely to define their victims as unrelated, often going so far as to identify themselves and their tribe as "human," and other social groups as nonhuman, subhuman, in short, as distantly related as possible. There is no known tribe or ethic group that prized the flesh of relatives as an especially yummy appetizer or as main course.

The preceding somewhat tongue-in-cheek description of the food chain is meant to show that whatever the importance of complex, interesting behavior such as altruism, a fundamental kin-based distinction goes deeper yet, permeating many aspects of human life. Creatures are especially unlikely to be altruistic toward one another when their stomachs are empty and their shared genes are few.

So, blood—that is, human blood—is not only thicker than water, it is also thicker than monkey blood, which is thicker than cow blood, which is thicker than fish blood, which is thicker than carrot blood, and so forth. (Unless one happens to be a monkey, or cow, or . . .)

Continuing up the pyramid of victimization and violence among human beings, notice that tribes—which usually means extended groups of relatives—"prey" on other tribes if not for food then for land and resources. And among human beings, physical traits show distinct commonalities, which leads to the concept of races and ethnic groups. To some degree, a race is a tribe writ large, a group of individuals with demonstrable physical attributes that display common ancestry.

No responsible member of the human community can deny the toll that racism and ethnic hatred have caused the human species.

This is the ugly underbelly of kin selection: not selfishness, but racism, a special form of intolerance toward others, those who are biologically different, or, if nothing else, who look that way. The average human being 10,000 years from now will probably look more international, more global, more racially indeterminate than most people today, as global communications, travel, and economic exchange make our shared planet a smaller place, with fewer isolated pockets of inbreeding, and more reproductive exchange between people. But for now, even with the great waves of migration and colonization and world-wide travel that characterized the end of the twentieth century, most of us still wear our genetic histories on our faces.

Although some people claim that the various human races are socially constructed and thus, biological fictions, the reality is otherwise. To be sure, there is no simple answer to the question, "How many races are there?" or "Are such-and-such a distinct race?" *And there is absolutely no doubt that all human beings are members of the same species.* But it is also clear that Caucasians, say, are easily recognized as distinctly different from Chinese, and that either group is different from black Africans. Moreover, there can be no doubt that such differences—superficial as they are—reflect genetic differences: After all, black parents produce black offspring, pink parents produce pink offspring, and so forth. A Martian zoologist, viewing the human species, would have to conclude that it is made up of distinct subspecies, varieties, races or whatever term is unquestioningly applied, for example, to recognizable breeds of cats, dogs, horses, and so on.

Let us be clear: This is not to say that "race" is a particularly meaningful characteristic, nor is there any way that the human races can be in any way ranked as better or worse, superior or inferior. Moreover, many—perhaps most—of the differences among the races are more apparent than real; there is more genetic diversity, for example, among black Africans than between Caucasians and Asiatics. Nonetheless, racial traits exist, just as eye color exists, along with earlobe shape or blood type, and at least some of the differences among the races result from differences in their genes. This recognition, although it may make some well-meaning people uncomfortable, is demanded by old-fashioned intellectual honesty.

Those physical traits that characterize the various human races are the relics of genetically isolated groups of people (tribes) who remained isolated for many generations. Australian aborigines evolved kinky hair, residents of the Mongolian steppe evolved eye-folds, and so forth. Geography was presumably the cause of this genetic isolation. Even as the races came increasingly into contact, interbreeding has been limited by cultural traditions that have generally kept individuals from marrying far outside their social/biological group.

What does this have to do with kin selection, or with racism? Just this. As we have seen, human beings—like other living things—are predisposed to behave benevolently toward close relatives over distant relatives, and to favor distant relatives over strangers, at least in part because the closer the relative the higher the probability that genes will be shared. When it comes to recognizing one's kin, it seems highly likely that physical similarity will be important: Everyone knows that relatives tend to resemble each other. And conversely, the less the resemblance, the less the likelihood of a close genetic relationship.

Skin color, eye shape, hair texture, physical size, nose shape, and other phenotypic differences among human beings reflect different ancestries. Even though all *Homo sapiens* share 99 percent of their total genes in common, just as all horses, for example, share 99 percent of their genes in common, the 1 percent that makes up physical or behavioral differences announces one's breeding or pedigree. The more differences, the more distant the genetic relationship. And the more differences, one can predict with some dismay, the less altruism.

The result may well be that human beings are naturally inclined—as a regrettable consequence of kin selection—to behave non-altruistically toward others whose physical traits mark them as truly Other, that is, unlikely to be closely related. Once again, since this issue is so fraught with emotion and the potential for misunderstanding, let's be as clear as possible: Racism is in no way rendered acceptable just because it may be, to some extent, "natural." To the contrary, it is unacceptable, a practical and moral wrong that human beings are obligated to struggle against. But the fight against racism is not abetted by ignorance as to its origin.

To some degree, the more obvious the signs of genetic differences, the more intense are ethnic wars and genocide. Large, white-skinned Europeans had little sympathy for small, dark-skinned Aborigines in Australia, or pygmies in Africa. To some degree, the tall, white Europeans with smallish noses favored the tall, dark-skinned people with smallish noses in Central Africa—the Tutsis—over their shorter, broader-nosed neighbors, the Hutus. Thus the Tutsis were given economic and political advantages over the Hutus, fueling some of the animosity that subsequently erupted in Central Africa. Furthermore, even without this push from European racism, Hutus and Tutsis—just like so many ethnically distinct groups—would doubtless have found justification for prejudice and violence; that is, for extreme forms of non-altruism.

In the musical *South Pacific*, a heroic (Caucasian) lieutenant falls in love with a Polynesian woman. Reacting angrily to the racism of his society, the lieutenant sings, "You've got to be taught, before its too late, before you are six or seven or eight, to hate all the people your relatives hate . . . You've got to be taught, to be afraid, of people whose eyes are oddly made, and people whose skin is a different shade. You've got to be carefully taught . . . " Racism undoubtedly *can* be taught, and regrettably, it often is. So, fortunately, can racial tolerance and compassion. The point is that to some extent—exactly how far is unknown—people may indeed *have to be taught* tolerance, because left to their own devices, the whispers of kin selected genes within most people seems to predispose them to a degree of bigotry that our species cannot afford.

Ironically, evidence is rapidly accumulating that human racial differences are literally no more than skin deep, that they reflect a very small proportion of the human genotype, probably less than 1/10 of 1 percent. Those racial categories that appear so prominent to so many people evidently reveal our tendency to establish social categories far more than they reflect biological reality. Nonetheless, human beings are acutely sensitive to the details of "exterior packaging" by which we identify each other as family, friends, or foe. It may be a tragic paradox that in unconscious pursuit of kin-selected benefits, we have come to exaggerate the significance of superficial differences that are just that: superficial.

The good news is that tolerance can be taught, because of the enormous behavioral plasticity and lifelong capacity for learning and adjustment that is our species' heritage. You *can* teach an old dog new tricks, and an old person, new insights. Recent developments in neurology have convincingly demonstrated that even in old age, the human brain possesses great capacity for compensation, even after strokes or injury. From childhood to senility, people can acquire new awareness along with new information.

We are children of the same mother—evolution—all of us nourished by the earth's good juices, yet our genes may well be programmed to see only narrower distinctions. To transcend ourselves, and our genes, is the uniquely human prerogative, as well as, increasingly, our responsibility.

Perhaps knowing the fundamental kinship among people—no matter their outward appearance—may help to dispel racist assumptions. If we all share 99 percent of our genes, why discriminate on the basis of a few trivial differences? The logical extension of the biological perspective is echoed in the Jain equivalent of the Golden Rule: "In happiness and suffering, and in joy and grief, we should regard all creatures as we regard our own self, and should therefore refrain from inflicting upon others such injury as would appear undesirable to us if inflicted upon ourselves."

Fortunately, racism and intolerance are not the whole story; the human repertoire also includes altruism, benevolence and love. Thanks to Darwin and his intellectual descendants, we can begin to understand how these inclinations—altruistic no less than violent—have come about.

Notes

1. In chapter 2, we considered the case of aggressive behavior among Japanese macaques, in which relatives are more often attacked, but only because they are more often encountered.
2. Napoleon Chagnon and Paul E. Bugos. 1979. Kin selection and conflict: An analysis of a Yanomamo ax fight. In Chagnon and Irons, eds. Evolutionary Biology and Human Social Behavior. Duxbury: North Scituate, MA.
3. N. Chagnon, 1975. Geneology, solidarity, and relatedness: Limits to local group size and patterns of fissioning in an expanding population. Yearbook of Physical Anthropology: 19: 95-110.
4. S. B. Johnson and R. C. Johnson. 1991. Support and conflict of kinsmen in Norse earldoms, Icelandic families and the English family, Ethology and Sociobiology 12: 211–230.
5. Austin L. Hughes. 1988. Ethology and Sociobiology 9: 29-44.
6. John M. McCullough and Elaine York Barton. 1991. Ethology and Sociobiology 12: 195-209.
7. James Acheson. 1988. The Lobster Gangs of Maine. University Press of New England: Hanover, NH.
8. S. Essock-Vitale and M. McGuire. 1985. Women's lives viewed from an evolutionary perspective. II. Patterns of helping. Ethology and Sociobiology 6: 155-173.
9. E. Burnstein, C. Crandall and S. Kitayama. 1994. Some neo-Darwinian decision roles for altruism: Weighing cues for inclusive fitness as a function of the biological importance of the decision. Journal of Personality and Social Psychology 67: 773-789.
10. D. P. Hogan and D. J. Eggebeen. 1995. Sources of emergency help and routine assistance in old age. Social Forces 73: 917-936.
11. N. L. Segal, S. M. Wilson, T. J. Bouchard and D. G. Gitlin. 1995. Comparative grief experiences of bereaved twins and other bereaved relatives. Personality and Individual Differences 18: 511–524.
12. William Jankowiak and Monique Diderich. 2000. Evolution and Human Behavior 21: 125-139.
13. John C. Loehlin. 1992. Genes and Environment in Personality Development. Sage Publications: Newbury Park, CA.
14. Nancy L. Segal. 1984. Cooperation, competition, and altruism within twin sets: A reappraisal. Ethology and Sociobiology 5: 163-177.
15. Natural selection and the analysis of human sociality. 1977. In Changing Scenes in the Natural Sciences. C. E. Goulden, ed. Academy of Natural Sceinces, Special Publication 12: 283-337.

16. 1931. Djuka: the Bush Negroes of Dutch Guiana. Viking Press: New York.
17. References in Mark Flinn. 1981. Uterine vs. agnatic kinship variability and associated cousin marriage preferences: An evolutionary biological analysis. In Natural Selection and Social Behavior. R. D. Alexander and D. Tinkle, eds. Chiron Press: New York.
18. S. J. C. Gaulin and A. Schlegel. 1980. Paternal confidence and paternal investment: A cross cultural test of a sociobiological hypothesis. Ethology and Sociobiology 1: 301–310.
19. Martin S. Smith, Bradley J. Kish, and Charles B. Crawford. 1987. Inheritance of wealth as human kin investment. Ethology and Sociobiology 8: 171–182.
20. R. L. Trivers and D. E. Willard 1973 Natural selection of parental ability to vary the sex ratio of offspring. Science 179: 90-92.
21. John Hartung. 1982. Polygyny and inheritance of wealth. Current Anthropology 23:1–12.
22. J. F. Hare and J. O. Murie. 1996. Ground squirrel sociality and the quest for the "holy grail": Does kinship influence behavioral discrimination by juvenile Columbian ground squirrels? Behavioral Ecology 7: 76-81.
23. Lixing Sun and Dietland Muller-Schwarze. 1997. Sibling recognition in the beaver: A field test for phenotype matching. Animal Behaviour 54: 493-502.
24. Group Psyuchology and the Analysis of the Ego. XVIII, Basic Books.
25. 1966 The Ego and the Mechanisms of Defense. Basic Books.
26. Irwin Tessman, 1995. Human altruism as a courtship display. Oikos 74: 157-158.
27. 1859 (1961) The Descent of Man and Selection in Relation to Sex. The Modern Library: New York
28. Ronald C. Johnson. 1996. Attributes of Carnegie Medalists performing acts of heroism and of the recipients of these acts. Ethology and Sociobiology 17: 355-362.
29. Cloninger, C. Robert. 1987. A systematic method for clinical description and classification of personality variants, Archives of General Psychiatry 44: 573–588.
30. Joseph Shepher. 1983. Incest: a biosocial view. Academic Press: New York.
31. Arthur Wolf. 1966. Childhood association, sexual attraction and the incest taboo. American Anthropologist 68: 883-898.
32. Jane Austen. (1814) 1986. Mansfield Park. Buccaneer Books: Cutchogue, NY.
33. Martin Daly and Margo Wilson. 1988. Homicide. Aldine de Gruyter: Hawthorne, NY.
34. Daly and Wilson, Homicide.
35. John Hartung. review of "A People That Shall Dwell Alone," in Ethology and Sociobiology 16: 335-342, 1995.

4

Reciprocity: Doing unto Others

Even if you aren't related to someone, you can still be an altruist, and with evolution's blessing. You can follow the Golden Rule: As Jesus said, "Whatsoever ye would that men should do to you, do ye even so to them." In other words, be nice, and beyond whatever rewards you may garner in heaven, maybe others will be nice in return, here on Earth. Biologists call it reciprocity.

Chapter 2 explained why natural selection favors dispensing aid toward genetic relatives, along with some animal examples. Chapter 3 described how this works for people. Now, we'll look at how evolution can select for altruism—still at the level of the gene—even when the altruist and beneficiary aren't genetically connected.

Requirements for Reciprocity

In a sense, reciprocity is plain common sense: "you scratch my back, I'll scratch yours," or "one good turn deserves another." Its echo resounds whenever we assess someone as "reliable," "trustworthy," or "somebody you can (or can't) count on," as well as in such laments as "what did I do to deserve that?" or the satisfying "he had it coming." But sometimes it takes a while for common sense to become scientifically legitimate. In biological thought, reciprocity is surprisingly new, first analyzed in detail by Robert L. Trivers (then at Harvard, now at Rutgers), in a landmark research paper published in 1971 and titled "The evolution of reciprocal altruism."

Here is the idea. Imagine you have some food and are about to eat it. Along comes someone else who begs for a portion. If you give in and share, there is less food for yourself. If your hungry importuner is a genetic relative, you should be generous—that is, altruistic—if the two of you are related, all the more so if you are closely related. (You can also be expected to be more generous if he or she needs the food, less so if you can use the extra bite yourself). But there is an additional set of circumstances in which it would pay your genes to donate food to the beggar, even if the two of you are not related at all.

There is one key provision: in the long run, the benefit of being altruistic has to overcome the cost. In other words, there must be a good chance that sometime in the future the tables will be turned—literally! If the recipient of your aid, for example, will someday have food when you don't, and if your good deed will be remembered and repaid in kind, then reciprocity may be a tasty feast for everyone involved.

It sounds simple enough, but actually these are demanding, rigorous conditions, which probably don't occur very often. The problem is this. Imagine that you have just given some food to a beggar, unrelated to you. As soon as you have done so, your evolutionary bank account has gone down by one notch, while the beggar's has gone up. This wouldn't be a problem in the long run, *if* you will eventually gain more than you lost by donating the food in the first place. Not impossible, since if you have just killed a large animal, you probably wouldn't require every single slice of meat. For the hungry beggar, by contrast, just one slice might mean the difference between survival and starvation. So, you lose a little; he gains a lot. If some time later, *you* are starving, and your beggar buddy has hit the jackpot, it could be a good trade all around: Give a little when you are fat and happy in return for a likely assist when you're really in need. In some ways, the transaction is equivalent to a healthy person banking some of her bone marrow, to be used later, if necessary, when the situation is dire. But when it comes to bone marrow, there is no question that the donor will be able to retrieve the investment. In the case of reciprocity, on the other hand, there is a huge and lingering doubt: the beggar/beneficiary might *not* pay you back.

Perhaps this is what Polonius, in Shakespeare's *Hamlet*, was driving at when he advised his son, "Neither a borrower nor a lender be."

Remember, after the first act of altruism, you (the altruist) have lost something—in our example, some food, which translates into some fitness—while the recipient has benefited. What is to stop him or her from profiting from your generosity, then snubbing you when payback time rolls around? It is more blessed, we are told, to give than to receive. But it is awfully tempting to take, and not to repay. (Maybe that's why giving is so "blessed"!)

In any event, this is the major difficulty that any system of reciprocal altruism must confront. Economists refer to it as the "free rider problem." Their classic example is a lighthouse: It is costly to build, but once constructed, everyone can benefit from its light. Hence, there is a temptation to let other—altruistic—people pay for its construction, and then to get a "free ride," taking advantage of the cost incurred by others. When it comes to lighthouses, the solution is to have the government do it; that is, to force everyone to bear the costs. When it comes to a would-be reciprocal exchange between two individuals, such solutions are more difficult to enforce.

There are also some minor problems, all of which must be resolved for reciprocity to work. For example, altruist and beneficiary must have a good chance of meeting again later. Otherwise, a debt incurred cannot be paid back. (Hence, reciprocity is less expected among total strangers, who are likely to melt away into an anonymous crowd.) There must also be a sufficient probability that at some time, the tables really will, in fact, be turned, and that the beneficiary will be in a position to reciprocate and repay the debt. (When giving food to a hungry colleague, better to favor someone who is, say, a competent hunter, just temporarily down on his luck, than a loser who is chronically empty-handed and thus a bad investment.) The beneficiary must also be able to recognize the altruist, and not to dispense repayment to others who don't deserve it. But above all looms the temptation to cheat.

Most people assume that cheating is nasty, unfair, maybe even despicable. And there is probably a good reason why this is so, why cheaters generally evoke such intolerance. In fact, it may be precisely *because* reciprocal altruism is so important to people—and at the same time, so vulnerable to cheaters—that the non-reciprocator is singled out for such criticism. "There is no duty more indispensable," wrote Cicero, "than that of returning a kindness. All men distrust one forgetful of a benefit."

Pity the poor cheater, however: He is doubly tempted. First, he gains by receiving something for nothing; then he gains yet again by ducking out when it is his turn to reciprocate (if he *does* reciprocate, then he would instantaneously be losing something, however small, just as the initial altruist did). Of course, the cheater may really be forgetful, as Cicero suggests, but it seems more likely that just as there can be Freudian slips, there are "Freudian forgets," or rather, biologically mediated inclinations to deny one's obligation. Maybe beggars can't be choosers; they can, however, be cheaters.

But as Trivers pointed out, the situation is not hopeless. Just as there are costs to reciprocating (hence, temptations to cheat), there are also some good reasons to do one's part and play by the rules.

Most important, if cheaters can be identified and then punished, this in itself would make cheating costly. Such punishment could be administered directly, as by a physical blow or some form of reprimand or indirectly, by excluding the cheater from subsequent exchanges of favors. Trivers suggested that the requirements for reciprocal altruism to work are so stringent, and, by the same token, the temptations to cheat are so great, that reciprocity might well be a human specialty. He even coined the phrase "moralistic aggression" for the peculiarly intense pressure that people exert upon suspected cheaters, in an effort to bring them into line and keep them there. Hell hath no fury like an "altruist" (actually, a would-be reciprocator) whose generosity is not repaid.

A Bird, a Bat, and a Baboon

That's the theory. Like kin selection, it is elegantly simple. Also like kin selection, it is powerful, although less widespread. For examples, we turn to three cases of reported reciprocal altruism among animals: a bird, a bat, and a baboon.

The bird is a lovely creature found in East Africa, the white-fronted bee-eater. These animals were studied over a period of several years by a team of researchers led by Stephen Emlen of Cornell University. Bee-eaters build their nests by excavating tunnels about one meter long in vertical cliffs along river banks. An active bee-eater nest consists of two breeding adults plus often—although not always—one or more additional adults who are "helpers." These helpers assist in excavating the nest, feeding the nestlings, defending and caring for the young. In some cases, they may be genetically related to the offspring being reared; if so, their behavior seems explicable in terms of kin selection. However, at least sometimes, helpers are unrelated to the individuals being helped.

When Emlen and his research team began investigating the social behavior of these animals, it was quite by accident that they discovered bee-eater helping. In the course of examining breeding birds on their nests, several pairs were disturbed. These individuals abandoned their nesting efforts, whereupon, to the surprise of the researchers, they joined other breeders, as helpers. It soon became clear that the course of bee-eater reproduction often runs into difficulty, even without biologists weighing, measuring, and observing. Under natural conditions, breeding is a chancy affair for these small birds. The rains on which they depend—to generate the large numbers of insects needed by the ravenous young—often don't materialize. And when they do, sometimes they are torrential, causing nests to be flooded and destroyed. Should this happen, or if nesting fails for any other reason, the unsuccessful would-be parents are likely to move in with another pair of successful breeders, and help out.

Not only that, but they don't donate their services at random. In Emlen's study, 85 percent of those who joined other breeders chose a pair with whom they had shared a social relationship within the past year or so; they were particularly predisposed if at least one of the breeders had helped them sometime earlier.[1]

Next, the bat. Specifically, that favorite of horror stories, the vampire. Vampire bats (technically, *Desmodus rotundus*) really do exist, although they are found only in Central and South America, not Transylvania. They don't actually suck blood; rather, they lick it, after making a tiny incision with their needle-like incisors. And their preferred "prey" is livestock—cattle and especially horses—not people. (There are two other species of vampire bat, the white-winged and hairy-legged vampires, which feed mostly on bird blood and about which very little is known.) In Central America, where they have been most carefully studied, vampire bats are reported to land on the mane or tail of their victims, although I have seen them in the cattle country of Colombia, landing on the ground nearby, and then tiptoeing incongruously, shoulders hunched high, apparently trying not to awaken their prey.

Belying their nasty reputations, vampires use a buddy system. They are, if you please, reciprocating altruists.

Vampire bats roost in caves or hollow trees, in groups of eight to twelve adult females, including their female offspring (young males leave home early). Gerald Wilkinson of the University of Maryland, who has studied these creatures for many years, finds that vampire daytime roosts are not made up of close relatives: The average genetic relatedness among female vampire bats is very low.[2] Within these roosts, however, females establish long-term associations, often lasting several years, and sometimes more than ten. (Despite their small size, incidentally, most bats have a long life span, which, in itself, seems to be favorable for reciprocity, probably because it increases the chances of eventual payback.)

Vampires share food. To put it frankly, they regurgitate blood to each other. A well-fed bat has a distended belly, easily visible to the human observer, and presumably even more obvious—and attractive— to a hungry bat. The solicitor licks under the potential donor's wing, then her lips; if successful, she gets a vomited meal of blood in return. This may sound less than delightful, but for a vampire bat, it is just terrific, the very stuff of life. Such sharing is important, since one-third of all vampire bats under two years of age come back from their nocturnal flights without having gotten a blood meal, generally because the cows or horses on which they attempted to feed woke up and brushed them away. Such wariness poses a serious problem for the bats, since they need from 50 percent to 100 percent of their weight in blood every day, and if they fail to obtain a blood meal for just two days running, they can starve to death. All vampire bats come back hungry on occasion, and it appears that individuals who are successful on one night fail, unpredictably, on another. So, the life of a vampire bat is one of constant alternation between success and failure, which satisfies that key requirement for reciprocal altruism: Recipients and donors trade places.

Wilkinson found that bats established relationships of reciprocating regurgitation, and furthermore, that such pairs were composed only of those that had been together at the roost at least 60 percent of the time. In addition—and as expected by theory—the cost of donating blood is less than the benefit of receiving it: A recently fed bat can save the life of its roost-mate, at relatively little cost to itself. Wilkinson also found that well-fed vampires direct their bounty toward others who are especially needy and close to starvation. And finally, bats that had been starving, and that received mouth-to-mouth transfusions, were likely to reciprocate the next time around, when their hunting had been good and the previous donors were in need. It is not clear whether there are any cheaters among vampire bats, but they evidently can recognize each other as individuals, so it seems likely that any cheaters could readily be made to suffer a dire penalty for non-reciprocation: being denied food when they need it. (It has also been found that female vampire bats do not have a monopoly on reciprocity; males share blood in the same way.)[3]

Food sharing—beyond parents helping out their offspring—is very rare among mammals generally. In addition to vampire bats, it has been observed only among African wild dogs, hyenas, chimpanzees, and human beings. But then, food-sharing isn't the only avenue for reciprocity.

For our last example we turn to baboons in East Africa. Baboons are not notable for their food-sharing. They do, however, help each other in a different way. Picture the scene: You are a mid-ranking male baboon, smitten with desire for a particular female. But here's the rub. Your lady-love is in the close company of another male. Moreover, he is an impressive fellow, socially dominant over yourself. You saunter over to a third male baboon, and by looking from him to your beloved's swain, you signal your request that number three join you in forming a temporary coalition to drive off your competitor. He complies and the two of you are successful, since even a baboon Rambo is no match for you and your henchman combined. Then, your buddy leaves, and you get to reap the reward; namely, to copulate with the female baboon of your dreams.

What is wrong with this scenario? From your prospective, nothing. With a little help from your friend, you and your genes have gained a fitness boost. Things are more problematic for your buddy, however. If he and you had been related, his altruism could be repaid via kin selection, somewhat like the case of brotherly assistance among wild turkeys. But male baboons leave their birth troop when they become sexually mature, so most adult males are only distant relatives, at best.

The answer appears to be that in the short term, there is no payoff at all to the baboon's buddy, just as there is no *immediate* benefit to a well-fed vampire bat who shares food with another who is starving. In fact, there is a cost, because aside from wasted time and energy, members of a baboon coalition can occasionally be wounded if the original male refuses to give up his consort without a fight. Careful observations of East African baboons by Craig Packer of the University of Minnesota have shown, however, that in the long-term, there is evolutionary method to the apparent madness of the helper baboon.[4]

Even if he doesn't get to mate with his buddy's female, "what goes around comes around." Sometime later, the altruist will likely solicit the assistance of the same male he had aided previously. And, being a good reciprocator, that male does his part, in effect, repaying his evolutionary debt, and keeping the system going.

That, at least, is the theory. The reality—as is often the case—has proven to be somewhat more complicated. Thus, it now seems likely that bee-eaters may gain directly (in terms of their personal reproductive success) as a result of their actions, at least on occasion. And baboons, it appears, are likely to compete for a lady baboon's sexual favors after they have chased off her original boyfriend.[5] Further research will doubtless clarify what's really going on . . . at least, somewhat.

For the present, however, bee-eaters, bats and baboons offer some of the clearest examples of reciprocity among animals. There are probably many others. Here is just a sampling: creatures as different as elephants, brown hyenas, and dwarf mongooses engage in so-called "allomothering" ("allo" = "other"), in which adults help to care for young not their own.[6] Although some of this is probably powered by kin selection and some by plain old-fashioned selfishness in disguise, it seems likely that reciprocity is also involved. When a captive bottlenose dolphin is about to give birth, she often teams up with another adult female, who serves as an "auntie," helping the newborn rise to the surface for its first breath, providing occasional babysitting, and even sometimes assisting with the birth itself.[7]

Among many species of whales, individuals are known to "stand by" when another animal is ill or harpooned; they even swim under stricken members of the group, keep them from sinking and appear to aid them in rising to the surface in order to breathe.[8] Supposedly, injured or tired human swimmers have also received such assistance. It is possible that such behavior—if true—results from erroneous allomothering by the animals involved.

Human beings, for their part, have made darker use of the allomothering inclinations of marine mammals. Sperm whalers used to capture and tether young whales, whose distress calls would draw others nearby so they could be killed.

Most of these cases cannot be pinned down to reciprocal altruism alone. If assistance is being rendered to genetic relatives (even distant relatives), kin selection could be involved. If there is a selfish compo-

nent—as, for instance, when individuals benefit personally by assisting others within their social group if they profit by being part of an
intact, well-functioning unit—then the behavior need not be strictly
altruistic. In reality, there is no reason why such behavior (or any
other, for that matter) should be unambiguously generated by any one
factor, acting alone. The living world is very complicated, and we are
just beginning to shine a few beams of light down its darker passage-
ways. Just because there are twists and turns doesn't mean that the
complexity of life can't be illuminated, at least a couple of feet at a
time.

Human Reciprocators

People are probably the world's champion reciprocators. It is even
possible that we owe our uniquely large brain to the demands of
reciprocation.

Consider this example. Someone is drowning. Without aid, there is
a 50 percent chance that he will die. If a potential rescuer comes to his
assistance, lets say there is a 5 percent chance that the rescuer will lose
his life in the process (in which case, both die). If the two individuals
are not related, then under a strict evolutionary regime, the rescue
should not take place, because it would impose a 5 percent cost on the
rescuer, and not return any benefit. But, lets add this: There is a good
chance that the tables will be turned sometime in the future, when the
rescuer may be drowning and his beneficiary able to render assistance.
Assume that once again, the drowning person would have a 50 percent
probability of dying, and the would-be rescuer, a 5 percent risk. Now,
by being part of a reciprocating system, both individuals would effectively substitute a 10 percent chance of drowning (5 percent plus 5
percent in each case), for the 50 percent liability that each would incur
if they had no potential reciprocator to help them. Even to the cold,
amoral calculus of natural selection—and even if the two individuals
share no genes whatsoever – it's a good deal.

Of course, the victim and rescuer could also be related, in which
case the motivation for assistance—and later, for reciprocation—would
be even greater.

People seem to be natural-born reciprocators. If the exchange is
immediate, no fancy evolutionary interpretation is needed: When we
purchase something, for example, there is the expectation of giving

and receiving equal value at the same time. It is presumed to be a mutually beneficial exchange, with neither party an altruist. But the possibility of one-sidedness arises whenever there is a time lag between the giving and the getting. Again, this one-sidedness works in more than one way: First, there is the one-sided, short-term benefit derived by the recipient, and second, the one-sided, short-term cost incurred by the altruist.

It is probably significant that whenever it comes to such exchanges, our emotions and sense of fairness are immediately engaged. One good turn deserves another: we all give at least lip service to this bromide, partly because it is hammered into us by our teachers, our families, our codes of morality and ethics, the constant prodding and insistent concern of society at large. Might there not also be a deeper, biologically based inclination as well? This inclination probably involves a complicated mix of tendencies, including—but not limited to—"reciprocate when necessary," and "cheat when possible." Maybe social rules wouldn't push so hard for reciprocity if people were naturally inclined to play fair most of the time, to return good for good without the urging of ethical principles and moral teachings.

Despite the tendency to cheat, or perhaps because of it, to share is to bond, worldwide. This is particularly true when it comes to food. Breaking bread, sharing salt, water or even a smoke: these are important symbolically as well as practically, helping to establish a personal, reciprocal relationship, not easily dissolved or denied, and sometimes even considered sacred. This is not only true of exotic cultures like Bedouin nomads or New Guinea highlanders. Receiving a gift, most Americans feel an obligation to send one in return. Think of your discomfort when you get something as inexpensive as a simple Christmas card, if you know that you hadn't sent one. Invited to someone's house for dinner, you can either reciprocate some time soon, or call the relationship off. Non-reciprocators, those who take and do not give, are quickly singled out as selfish undesirables, people one had better not associate with. (How many people will keep sending Christmas cards to someone who doesn't send one in return, or will keep inviting somebody over for dinner if they aren't invited back?)

We all enter into reciprocal relationships, pretty much every time we interact with other people. Of these, perhaps the most crucial is marriage. There is something hard-hearted and cold-eyed about seeing marriage as based on reciprocating exchanges rather than romantic love, but the likelihood is that much of what smoothes the interaction between two unrelated adults, living together in intimate association, is the expectation of reciprocity. Sometimes it is straightforward and not objectionable at all: You love me, and I'll love you. This includes, but is not limited to: you care for me and I'll care for you, you provide affection to me and I'll do the same for you.

Often, it is comparatively overt and direct: You pick up the kids, I'll walk the dog. You cook, I'll do the dishes. Other times it is covert and sometimes even unconscious: You provide money, I'll provide sex. You tolerate my kids (from a previous marriage), and I'll do the same with yours. I'll accept your binge drinking if you look the other way at my spending sprees. I'll bring home a paycheck, and put up with your emotional abuse as long as you agree not to abandon me and continue taking care of the kids. I'll tolerate your numerous girlfriends as long as we can have a waterfront home. I'll have a child with you as long as that entitles me to a piece of your family's wealth.

People are often ashamed and resistant to have their true motivations reduced to these dry equations. Especially when it comes to explaining why they stay in painful relationships, they would much rather talk of love or commitment, than of greed, self-aggrandizement, or reciprocity. Yet with rare exceptions, people, as other creatures, give in order to receive. In this sense, "love" itself may be little more than a rush of hormones, like heat in an animal with estrus cycles, plus an unconscious calculation of personal benefit: that the "beloved" has something major to offer, usually sex and money or power. In evolutionary terms, these are all important if they lead to enhanced fitness. From a biological perspective, "marriage" is a social contract for reciprocity in the form of mutual aid, possible reproduction, and the acquisition and protection of resources.

Social contracts of one sort or another operate in our most intimate, day-to-day lives, and when all parties abide by the expectation of reciprocity, everyone is far more likely to be satisfied than if someone assumes that he or she is entitled to "get" without "giving." With marriage, adults come together to share a household and perhaps reproduce. Each individual has strengths and weaknesses; each has resources to share, and needs to be met. In a healthy family, power and dominance are much less important than is equitable reciprocity. In fact, families characterized by a high degree of dominance and over-control are more likely to have emotionally ill children than are fami-

lies with better boundaries and less criticism: in short, the more reciprocal love, the more health and happiness.

The likelihood is that reciprocity contributes not only to mental
balance and social harmony but to a healthy sense of fairness. Saying
that a relationship is characterized by a high degree of reciprocity is
another way of saying that it is fair and balanced. Putting it another
way: People may owe their sense of fairness to their recognition of
what it takes to maintain a smoothly functioning, reciprocal system.
Just as sugar tastes sweet, it "tastes" good to give a little to get a little;
or to get a lot if you've given a lot. Any way you slice it, however,
being a sucker is bitter indeed.

Maybe it tastes even sweeter to get a lot if you've given a little, but
this is rare, and moreover, most people may even harbor a deep personal suspicion of such asymmetry. ("If the deal seems too good to be
true, it probably is.") People are generally even more suspicious—and
usually for good reason—of those who give a lot, claiming that they
really aren't expecting to get back in proportion. Thus, when a wealthy
corporate executive or union leader contributes mega-dollars to a poli-
tician, and then claims that he or she is *not* buying special access or
favors, the public has every reason to be skeptical.

Like language, reciprocal altruism is clearly learned, and it follows
rules that are handed down by local tradition. But just as all human
beings use language, and all languages have certain "deep structures"
in common, all people engage in patterns of reciprocity, patterns that
are shared world-wide and that everyone—at some level—can understand.

"There is some benevolence, however small," wrote David Hume
in 1750, " . . . some particle of the dove kneaded into our frame, along
with the elements of the wolf and serpent."[9] That "particle of the
dove," responsible for our most benevolent, altruistic inclinations, is
that same part of human beings—and, incidentally, wolves and ser-
pents too—that responds to the adaptive call of either kin selection or
reciprocity—or both.

What are Friends For?

In theory, reciprocal altruism is quite different from kin selection. It
could even operate between individuals who are members of different
species, since at the genetic level, reciprocity is promoted by paybacks

that benefit genes within the body of the individual in question rather than through relatives. (For this reason, reciprocity is a better term than reciprocal altruism, since it refers to a process that is actually selfish, operating at the level of the individual as well as his or her genes.)

The fact that reciprocity does not require genetic closeness even suggests a new way of looking at domesticated animals: perhaps they have been selected as much for reciprocating with *Homo sapiens* as for deferring to them. For a pat on the head, or the pure joy of collaboration (not to mention a reliably provisioned food bowl), my Labrador will jump hurdles, ford rivers, and plow through deep snow to retrieve a stick. Traditional theory states that domestic animals have been bred to be subordinate, and to obey because human beings dominate them. Alternatively, maybe a crucial part of domestication is selection for being an eager reciprocator. People with pets generally provide social approval plus room and board, but neither is "free." They expect their dogs, birds, horses (but not cats!) to do their bidding, and moreover, to do so with apparent eagerness, not grim resignation as might be expected if mere subordination were the mechanism.

It is also possible, on the other hand, that the human penchant for pets may have been encouraged by a unique, benevolent perversion of kin selection. Thus, it is clear that the domestication of animals involved artificial selection, and whether this favored subordination or eager reciprocation, it may also have included a tendency—perhaps unconscious—to breed animals that mimic many of the features also provided by human babies. If so, then to some degree, pets may be like "nest parasites" such as European cuckoos, which are not just tolerated, but preferred, by their foster parents, and which ultimately benefit the pet/parasite rather than the host.[10]

In any event, and even though theory does not require it, most cases of reciprocity involve members of the same species. In practice, kin selection and reciprocity often blend together; although reciprocators don't have to be kin, there is no reason why they should not be. Having said this, however, let's attempt to keep the two concepts separate, if only because it helps simplify things.

A large proportion of human social behavior might well be encompassed and, in a sense, explained by adding together the combined effects of kin selection and reciprocity. With whom, for example, do people interact? Offspring, mates, friends, and those with whom we do business or exchange information. Of these, the first relates to inclusive fitness, whereas marriage represents a mixed case (because it entails, in addition to the reciprocity described above, shared interest in individuals—children—with whom husband and wife both share genes, although different ones). The last two—friends and business associates—fall within the realm of reciprocity.

Pure kin selected altruism is probably rare: most people generally expect some kind of reciprocity even from relatives (including many interactions with offspring). The common advice to avoid doing business with relatives could stem from the fact that relatives may feel less than the usual need to act reciprocally, as well as from the social complications of cheating among family members.

Although pure kin selection, without reciprocity, may be infrequent, pure reciprocity without kin selection is comparatively common among human beings. It may even be what friendship is all about. "The only reward of virtue is virtue," wrote Ralph Waldo Emerson. "The only way to have a friend is to be one." We count on our "friends." And they count on us. Put more strongly, we identify someone as a friend if, and often only if, we *can* count on him or her. What do we count on friends *for*? For neither more nor less than reciprocity: for the reliable, useful and pleasurable exchange of favors, assistance, valued company, sympathy, shared interests, and so on. Think of how often, when a friend does someone a good turn, the phrase pops up: "That's what friends are *for*."

This perception is no less valid for seeming cynical. When it comes to business associates and others with whom we interact but do not strictly consider friends, relationships are governed even more objectively and impersonally by the expectation of getting fair value in return for our offerings. We exchange goods or favors, for example, with the shared expectation that each side is being fairly recompensed. Even a simple request for information—such as the time of day—is repaid with a "thank you," which acknowledges the small debt, and also, in a sense, confers a degree of status upon the donor.

Competitive Altruism?

What seems to be altruism may be subtly competitive as well, in animals no less than human beings. Chimpanzees, for example, sometimes "give" meat in exchange for social control, which increases the rank of the donor.[11] The result is a pattern of dubious donations, in which individuals benefit themselves by giving, and others debase themselves by accepting.

It has recently been confirmed that chimpanzees actively hunt for meat, especially choosing small colobus monkeys as their victims. Interestingly, the most reliable predictor of whether a chimpanzee troop will initiate such a hunt is whether there is an estrus female within the troop. If so, and if the hunt is successful, then chimpanzee males will present some of the meat to the female, in return for sex.[12]

Israeli zoologist Amotz Zahavi made some intriguing observations in his long-term study of a species of bird known as Arabian babblers, animals in which "helping at the nest" is common. In the seemingly cooperative and altruistic world of Arabian babblers, young adults frequently bring food to the nestlings and also serve as anti-predator sentinels. Although this behavior might seem explicable by kin selection, or possibly even reciprocity, Zahavi found, curiously enough, that the dominant, breeding adults do not welcome such "help." They try to prevent it. Zahavi's explanation—controversial, but worth considering—is that the "helpers" are really seeking to help themselves, by acquiring the additional social prestige that comes from bestowing aid on others.[13]

There is a curious anthropological parallel to this behavior, the so-called "potlach" ceremony of Northwest native Americans such as the Kwakiutl of Vancouver Island. High-ranking Kwakiutl used to engage in competitive show-off contests, in which they sought to elevate themselves and diminish the prestige of others by intentionally destroying as much of their own property as possible. In less exotic circumstances, most people understand that there is something demeaning about accepting favors from others: When John D. Rockefeller famously scattered dimes to the crowd of admirers who flocked around him, he italicized the difference between himself and the hoi polloi. Even though his beneficiaries ended up somewhat richer for the experience, Rockefeller, as benefactor, emphasized that he was above them.

Unless they are in dire need, or unless, as in Rockefeller's case, the donor is so exalted that no question of competition arises, people around the world are often hesitant to accept gifts. This may well be because deeply ingrained in the human experience is a recognition that nearly always, gifts come with "strings attached." They tend to be entangling however they are interpreted. Either they carry an implicit expectation of reciprocity, or, if they do not, then to accept gifts without future reciprocity is to be on the receiving end not only of the obvious "goods" in question, but also of some less obvious "bads" that come with a social—and implicitly, a biological—put-down.

Even the well-known and oft-derided reluctance of men to ask directions may be related to this phenomenon of subtle altruistic/competitive one-upmanship. To ask for directions is, in a sense, to request a hand-out, to place the direction-giver in the superior position of donor and the asker, as lowly recipient of another's largesse. Men are, to a large extent, the more aggressive and competitive sex; small wonder, therefore, that they often shy away from placing themselves at any social disadvantage, even one that appears so trivial.[14]

Either/Or versus Both/And

Law professor John Beckstrom of Northwestern University has shown that the expectations of reciprocity figure prominently in U.S. judicial opinions, even when the judges involved make no reference whatever to reciprocity, evolution, and so forth. As Beckstrom points out, the judgments of judges can be seen as reflecting the received wisdom of society's "elders." He reviewed 400 legal proceedings from the 1850s to the 1980s, each of this general sort: Individual A donates goods or performs services for individual B, then subsequently brings a lawsuit against B for refusing to pay. Individuals A and B often disagree about whether a verbal agreement specified repayment.

Beckstrom found that several principles have guided judges faced with rendering a decision in such cases. First, there is no free lunch: it is assumed that when someone provides goods or services for someone else, the donor is entitled to repayment. In other words, reciprocity is implicit whenever people interact such that something is given and received. Second, there are two important exceptions to this presumption of reciprocity: (1) when the donor and recipient live in the same household or (2) when they are close relatives.

In the first case, legal thinking assumes that co-householders are already engaged in an ongoing reciprocal relationship. Hence, payback is not automatically required, because A and B are probably involved in a continuing process of exchange. As to the second exception, kin selection speaks for itself. It speaks rather loudly, in fact, even for judges who have never heard of evolutionary genetics. Thus,

Beckstrom found a remarkable consistency in legal opinions: The more distant the genetic relationship between disputants A and B, the greater the expectation of payback. Another way of looking at it: the closer the relationship, the less need for return payment. As one judge's opinion put it, "While . . . a presumption of gratuitous service extends to all relatives . . . it grows weaker and becomes more easily rebutted as the relationship recedes." In short, "gratuitous" services—those offered for free—are to be expected among close relatives, less so among more distant relatives, and when it comes to non-relatives, there is no such presumption at all. When two individuals are genetically related, in short, this in itself justifies benevolence. And so, no reciprocation is required when favors are directed to close relatives.[15]

One of the interesting complications revealed by Beckstrom's research is the continuing difficulty of separating kin relations from reciprocity. Thus, people living in the same household are likely to be genetically related and to be reciprocators. So it probably isn't either kin selection or reciprocity, but both, to varying degrees.

An intriguing bit of research has helped distinguish kin selection from reciprocity, at least among animals. It was conducted by the husband and wife team of Robert Seyfarth and Dorothy Cheney, both of the University of Pennsylvania. Seyfarth and Cheney have spent many years studying vervet monkeys in Amboseli National Park, Kenya. In one study, they identified pairs of monkeys that had been associated with each other. Some were close relatives; others, distantly related, if at all. Some had been busily grooming each other for at least one minute; others had not. Then, a short time later, when the various individuals had separated, one member of each pair was played a recording of the other vocalizing, in a way that typically requests support in an aggressive dispute. Seyfarth and Cheney examined whether the monkey who heard the recording acted in a way that indicated willingness to come to the aid of the other. Here is what they found. If the pairs were close kin, it didn't matter whether they had recently groomed one another. Every subject responded strongly to calls for help. If the pairs were not closely related, however, then their immediately preceding behavior mattered a great deal. If mutual grooming had just taken place, willingness to help was strong. If not, then it wasn't.[16]

Thus, for closely related vervets, blood is thicker than recent reciprocation, while for unrelated monkeys, the question arises: "What have you done for me lately?" Similar considerations apply to food sharing among chimpanzees. For them, interestingly, being nice is often a matter of selfish tactics in disguise. There is a method to the beneficent "madness" of chimps who give food away. As primatologist Frans De Waal puts it, "sharing is no free-for-all." Examining more than 5,000 food transfers among chimpanzees housed at the Yerkes Regional Primate Research Center, in Georgia, De Waal and his associates found that

> the number of transfers in each direction was related to the number in the opposite direction; that is, if A shared a lot with B, B generally shared a lot with A, and if A shared little with C, C also shared little with A.

De Waal also reports, interestingly, that as with the vervet monkeys studied by Seyfarth and Cheney, chimpanzees don't insist that reciprocation always take place in the same currency. For example, "A's chances for getting food from B improved if A had groomed B earlier that day."[17]

When it comes to motivating kin selection, shared genes carry their own reward. But when reciprocal altruism is at issue, crucial questions arise, such as "Do I owe a payback?" "Does this individual owe me something?" and "Will he—or she—repay the debt?" Returning to the example of a drowning man: what is to stop him from receiving assistance when he needs it, but then, when the life-saver is in distress, failing to reciprocate?

As Robert Trivers, guru of reciprocal altruism, pointed out, systems of reciprocity may be inherently unstable, relying on a variety of psychological and social mechanisms in order to keep them going. Diverse psychological traits and social teachings may in fact owe their existence to the pressures of reciprocal altruism. The Golden Rule, for example, is a statement of idealized reciprocity: Do unto others as you would have others do unto you. Similar injunctions have been identified for most of the world's religions and ethical systems. They are especially important since as we have seen, when those others are truly "other"—that is, unrelated—there is a powerful, yet subtle pressure to behave otherwise.

The Prisoner's Dilemma

The slide toward selfishness has many representations, both in the popular imagination and real life. Perhaps it is what Carl Jung termed the "dark side," popularized in the *Star Wars* movies, although in most cases its practitioners are not as unabashedly evil as Darth Vader. Biological selfishness may be the evolutionary equivalent of the cartoonist's devil perched on one's shoulder and urging sin, or it may be comparatively passive, the temptation to cheat, to accept altruism and then refrain from reciprocating.

Or the temptation may be uglier yet, a decision to exploit the altruistic tendencies of others with a deliberate plan to take advantage of the altruist. Think, for example, of serial killer Ted Bundy, who walked along the University of Washington campus on crutches, or with his arm in a sling, appearing disabled, so that he could solicit help from attractive young coeds before abducting and murdering them.

The problem of cooperation—knowing when to cooperate and when to be self-serving—is an ancient conundrum, known as the Prisoner's Dilemma. Tons of ink have been spilled in efforts to explain (and get around) the Prisoner's Dilemma; it has intrigued mathematicians, economists, psychologists, legal scholars, political scientists, and—most recently—biologists.

Here is the Prisoner's Dilemma in skeleton outline. Imagine two prisoners, apprehended by the police and accused of collaborating in a bank robbery. The prosecutor would like to convict them both, but does not have enough evidence; so, his goal is to get them both to plead guilty, which isn't easy, since the prisoners know that the prosecutor's case is weak. But the prosecutor devises a devilish scheme. Trying to get them both to plead guilty, he separates the two—keeping each in a different cell and not allowing any communication between them—and makes each prisoner the same offer: "If you and your partner insist on pleading not guilty, you will both be convicted on a lesser charge, say, illegal possession of a weapon, and each of you will be sentenced to a year in jail. If both of you plead guilty, then you will each face a more severe sentence, five years in jail. On the other hand, if one of you pleads guilty to the robbery charge, thereby implicating the other, then I'll have enough evidence to convict the other guy, and the government will reward you for your assistance by letting you go free while your partner will be sentenced to forty years behind bars. But if you do not 'cop a plea' and your partner does, then you get the forty years and he goes free."

Under these conditions, each prisoner faces a cruel dilemma. Collectively, both would do best—just one year in jail—if they cooperated with each other and stuck to their plea of not guilty. However, each one fears what the other might do. Let's listen in on prisoner A trying to decide what to do: "B can choose, like myself, to plead guilty—that is, to be selfish—or stick with our cooperative claim of not guilty (be altruistic, hoping that I will do the same). If he pleads guilty (is selfish), then I had better do the same, since in that case if I choose to be altruistic I'd be a sucker and get a rotten payoff (forty years in jail), while B gets the highest return; namely, to go free. So in case he decides to be selfish, I had better be selfish too. But what if B chooses to be altruistic (that is, to plead not guilty)? In that case, of course I could be altruistic and plead not guilty too. I would then get a pretty good payoff: Exactly the same that he gets, since the two of us would be cooperating in a sense, and we'd both get off with just a year's sentence. But wait a minute! If B chooses to be an altruist, then I have a better choice yet. I could be selfish and plead guilty, in which case I get to take advantage of him—just as I had feared that he might take advantage of me—and I get the highest return (go free), while he gets suckered (40 years). Either way, therefore, no matter what B does, I protect myself and come out best by being selfish."

A's self-serving logic is impeccable—although perhaps immoral—and we can expect that B, reasoning the same way, would come up with the same answer. As a result, both players find themselves compelled by the logic of the "game" to be selfish; that is, to plead guilty. And this, if you remember, was the prosecutor's original aim. Here, then, is the dilemma: They both get lower payoffs than either would have gotten had they only been able to cooperate and behave altruistically (and trustingly) toward the other.

The Prisoner's Dilemma has been considered a simplified model for many things in modern life: nuclear disarmament, for example, in which each side fears to be nice (disarm), because it may be suckered if the other is nasty (keeps its weapons or builds more). So both wind up being nasty, stuck in an arms race, to the disadvantage of all concerned. The U.S. and the USSR were long mired in exactly this system, with both sides receiving a negative payoff, measured as the costs of hostility, distrust, wasted national treasure, and unmet domestic needs.

Although the Cold War is over, many people are still fighting cold wars in our personal lives, prisoners of their own, private dilemmas. There are tugs-of-war between spouses, between parents and children, worker and boss, between neighbors, in which each side would love to let go, but fears that it cannot afford to do so because the results would be embarrassing if not downright disastrous. And so, they settle for an ongoing stand-off, in which each side continues to struggle and pull: to be, in a sense, nasty, although it feels more like being self-protective. Backs aching, hands bloody and raw, each receives the mutual punishment of a stiff-necked (but thoroughly logical) commitment to "winning," as each defines the other as its enemy.

Fortunately, there are ways out. In the movie *WarGame*, a computer advises that in the case of nuclear weapons the only way to win is not to play the game. When it comes to tug-of-war, the easiest way to win is to let go. And even when it comes to Prisoner's Dilemma, the seemingly irrefutable logic of eternal, mutually punishing nastiness can in fact be side-stepped. Mathematically inclined students of the Prisoner's Dilemma have confirmed that nice guys don't necessarily have to finish last, especially if the Dilemma in question is a continuing, repeated series of interactions rather than just a one-shot deal.[18] When the possibility exists for repeated interactions, when—as one theoretician has put it—the "shadow of the future is long," it can actually pay to be nice, since nice guys can set up reciprocating partnerships and avoid the costs of everyone being nasty and inflicting pain on each other. It might be useful to move the apostrophe and to rewrite Prisoner's Dilemma as Prisoners' Dilemma, emphasizing that it is a shared problem that requires a shared solution.

It appears that the only "winning" strategy must include forgiveness. (This word belongs in quotation marks because it doesn't really win in the conventional sense; rather, it cannot be exceeded by any other. At the same time, it never does better than its "opponent.") Even after a treacherous opponent has recently double-crossed you, the strategy that is ultimately most successful is to cooperate—that is, to be forgiving of his or her past trespasses—so long as the opponent switches back to cooperation. The surest sign of repentance is benevolent cooperation, and the surest "winning" response is gracious acceptance of the other's apology.

At least in computer models, the take-home message of the Prisoner's Dilemma can be summed up succinctly. If your relationship is to be brief, without a likelihood of future interactions, and if you are presented with a situation that is potentially costly to you without much benefit, then think selfishly and take care of yourself. But if you are in a relationship that might be long term, be initially nice, then follow the lead of the other party. If he/she is nice, be nice. If nasty, be nasty. Don't hold a grudge. Don't be nasty if the other person has been nice. Be prepared to retaliate if the other has been nasty; don't be a sucker. Be aware of possible deception, but don't be paranoid. Or, most simply: cooperate, but don't be a fool.

Some Subtle Consequences of Reciprocity

The Prisoner's Dilemma is not a perfect model for why reciprocity is difficult. It makes a number of unrealistic assumptions: that the interaction between any two players only occurs once, that there are only a very limited number of possible responses ("reciprocate" or "cheat"), and that each player has no information as to what the other is likely to do, to mention just a few. However, Prisoner's Dilemma is a useful tool because it clarifies some of the problems of reciprocity: its vulnerability to cheating, the importance of judging a cheater by his or her actions, and of adjusting one's behavior accordingly.

When cheating is obvious, retaliation can be an equally obvious response, at least among species intelligent enough to recognize the transgression. Human beings are doubtless the champions at blaming one another for what they have or have not done, whether their behavior was warranted given previous events, and so forth. Punishment for non-reciprocation has thus far been reported for only one nonhuman animal species, and it should surprise no one that this animal is the highly intelligent chimpanzee. Here is an account by Frans De Waal, describing a complex encounter among captive chimps at the Arnhem Zoo in Holland:

> A high-ranking female, Puist, took the trouble and risk to help her male friend, Luit, chase off a rival, Nikkie. Nikkie, however, had a habit after major confrontations of singling out and cornering allies of his rivals, to punish them. This time Nikkie displayed at Puist shortly after he had been attacked. Puist turned to Luit, stretching out her hand in search of support. But Luit did not lift a finger to protect her. Immediately after Nikkie had left the scene, Puist turned on Luit, barking furiously. She chased him across the enclosure and even pummeled him.[19]

It is easy to imagine Puist saying to Luit: "I helped you. Why didn't you help me?" It seems that the wisdom of "One good turn deserves another" is no more lost on chimpanzees than on human beings. Part of this wisdom, moreover, could well derive from the penalty of being caught refusing to pay back the good turn that one has received.

For another example, De Waal compares the behavior of two adult female chimpanzees, Gwinnie and Mai:

> If Gwinnie obtained [food] . . . she would take it to the top of a climbing frame, where it could easily be monopolized. Except for her offspring, few others managed to get anything. Mai, in contrast, shared readily and was typically surrounded by a cluster of beggars.

Later, Gwinnie, with her stingy personality, encountered more resistance and threats when she begged food from the others. De Waal concludes, "It is as if the other apes are telling Gwinnie, 'You never share with us, why should we share with you!'"

On the other hand, cheating need not be 100 percent, as in a simple refusal to pay back one's debts; it could be camouflaged, in which incomplete reciprocity is offered. Perhaps instead of automatically jumping into the water to save someone who had saved your life, you did so only if conditions were especially safe. Or you shared food with someone who previously had shared with you, but you held back some of it. For the initial altruist who is currently in need, even a half-hearted, begrudging, and inadequate payback would probably be better than none at all, and the subtle cheater would also gain by this kind of deviousness. At the same time, any temptation to cheat—and any tendency to do so partially—could put a premium on the ability of others to detect such tendencies in potential partners.

The resulting balance is likely to be delicate, and fascinating. On the one hand, subtle and successful cheating would be favored. But on the other, the more it occurs, the stronger would be the evolutionary pressure to detect it. In the case of the male baboon whose assistance was solicited by a love-stricken colleague, it is pretty obvious whether or not assistance is rendered. Similarly for a hungry vampire bat: a well-fed animal, belly gorged with a recent meal, could not very well offer up only a few precious drops of horse-blood and then somehow claim that it had been generous. Not so, however, for certain other behavior, such as cooperative hunting or group defense, among lions or human beings for example. In such cases, when several individuals are expected to cooperate in bringing down large and potentially dangerous prey, the social rules may call for reciprocal altruism: each individual is expected to "pull its load," to run some risks and do its part, in return for which it gets a share of the forthcoming meal. (Not to mention just being allowed to remain part of the group.) But what is to prevent a hunter from running less than all-out, or from subtly protecting itself when push comes to shove and a downed wildebeest is wildly thrashing about with horns and hooves that could tear open an unguarded abdomen?

The result, once again: a finely tuned sensitivity to each other's behavior, and perhaps even the rudiments of a sense of justice. Indignation, too: the response to someone who takes more than he or she gives. It is more blessed, we say, to give than to receive, and unfair to receive more than you give.

Even systems of exquisite moral generosity are typically buttressed by an expectation of quid pro quo, the assumption—whether implicit or explicit—that we shall reap as we have sown. Even "turning the other cheek," the ultimate in non-reciprocated benevolence, is justified in Christian tradition, at least in part, by the promise of an eventual payback, albeit in heaven. And Western religious institutions, which consider themselves located more in the City of God than the City of Man, will nonetheless tithe, and/or pass the hat, assuming quite blithely that along with the receipt of spiritual benefits, parishioners also accept the responsibility to reciprocate in the coin of Caesar.

A similar pattern has persisted for more than 2,000 years in the Buddhist tradition. Thus, Buddhist monks

> live entirely in an economy of gifts. Lay supporters provide gifts of material requisites for the monastics, while the monastics provide their supporters with the gift of the teaching. Ideally—and to a great extent in actual practice—this is an exchange that comes from the heart, something totally voluntary.[20]

Whereas this "economy of gifts" depends upon purity of heart on the part of donors and recipients, and there is no explicit price for the teaching—not even a "suggested donation"—compassion alone is not the sole criterion for the exchange.

Thus the Buddha himself, no stranger to disinterested compassion, is reported to have said to his monks,

> householders are very helpful to you, as they provide you with the requisites of robes, alms-food, lodgings, and medicine. And you, monks, are very helpful to householders, as you teach them the dharma . . . In this way, the holy life is lived in mutual dependence, for the purpose of crossing over the flood, for making a right end to suffering and stress.[21]

Deception, Self-Deception, and Social Payoffs

Buddhist and Hindu holy men, with their vows of poverty and their ever-present begging bowls, are seen by many as the epitome of spiritual generosity. Yet, even they are expected to function within a tightly established norm of rules governing the exchange of godliness for alms. Among Buddhists and Hindus, there are rules against asking for donations at inappropriate times, against exaggerating one's spiritual attainments in order to garner a larger donation, even against using the rice in one's bowl to cover up especially yummy bits of food so as to make oneself seem more needy and thus garner a more generous donation!

Even the most meritorious, it seems, are sometimes tempted to cheat.

Cheating, interestingly, is not simply a matter of deceiving others. It could also set the stage for *self*-deception. Because people dislike cheaters, it is useful to maintain a public facade of morality and beneficence, of disinterested altruism. Cheating therefore does best when it is disguised, even to the cheater himself. (Or herself; don't lose sight of the androgynous nature of most of these generalizations.) Just as the best liar is the one who actually believes her falsehoods, the best cheater would be one who honestly believes that her actions are blameless, even laudatory. Such a person would be unlikely to give herself away, to betray her deception by a blush, a stammer, a faltering hand or reluctance to make eye contact. The best cheaters and deceivers are those who are blithely unaware of the true nature of their actions, and so, the likelihood is that natural selection perversely favors individuals who are oblivious to their underlying motivations.[22]

Like so much in revolutionary biology, this represents a departure from conventional wisdom, which generally holds that self-deception is a bad thing that eventually will rebound to the disadvantage of the victim (never mind that the victim is one's self). The key is to recognize that evolution has not acted to maximize the human grasp of truth, but rather, to maximize the fitness of human genes. Thus, self-deception may well have evolved as a means of deceiving *other* people and their genes, so it is irrelevant whether it operates via deceiving one's self as well. Any strategy that is successful is an evolutionary triumph, even though perhaps a moral disaster.

One result of all this strategizing and counter-strategizing is a kind of unending race, a positive feedback loop that intensifies with time. Mechanisms of detection become more subtle, because they are selected for, with savvy detectors doing better than trusting chumps. But at the same time, mechanisms of deception and self-deception also improve, because they, too, will be honed yet sharper by greater reproductive success on the part of those individuals, and their genes, capable of getting away with sneakiness. And of course, as detectors get more astute, this puts yet more premium on being a good deceiver . . . and vice versa, ad infinitum.

Don't get the impression, however, that reciprocity leads only to such dreary stuff as deception, cheating, and self-deception, along with efforts at detecting unsavory acts of this sort. There is also gratitude, sympathy, even conscience. "Gratitude," for example, is a positive response to another's altruism, a kind of unwritten receipt whereby the beneficiary acknowledges the benefit in a way that suggests that it has not only been noticed, but is likely to be repaid. And "sympathy"—a sensitivity to the plight of someone else, typically someone who is needy—emerges as a means of increasing the likelihood of altruism in the first place.

The plot thickens. Most people find a capacity for sympathy and gratitude attractive in someone else, not surprising if we unconsciously assess others by whether they are likely to be suitable reciprocators. As we have seen, this leads in turn to the whole question of friendship: what it means, what generates it, and what is likely to signal trouble. Few people choose to befriend someone who is nasty, lazy, selfish, unreliable, and so forth. What sort of people *are* we likely to like? Those we can count on, who will be if not altruistic, then at least dependable and conscientious, in short, likely to reciprocate. One of the quickest ways to sour a friendship is to let the friend "down." (Recall the earlier discussion of "what are friends for.")

As to conscience, Richard Alexander calls it as "the still, small voice that tells us how far we can go without incurring intolerable risks." He goes farther: "It tells us not to avoid cheating but how we can cheat socially without being caught."[23] call it evolutionary Macchiavellionism.

Then there is guilt, that insistent part of our conscience that lets us know when we have transgressed, and probably keeps us from going too far, from cheating too much. Fear of one's guilty conscience can even keep people from transgressing in the first place, which is especially useful, particularly if one is likely to be found out. When Hamlet complained that "conscience doth make cowards of us all," he was overdoing it. Conscience keeps us from being too selfish for our own good.

Not surprisingly, people are acutely sensitive to each other's underlying motivations, favorably disposed toward "genuine," uncalculating altruism and turned off by those who always seem to be "keeping score." (This doesn't mean, however, that they don't keep score themselves, just that they don't like others to do so, or to be transparent about it, themselves.) As we saw when examining marriage from the perspective of reciprocity, there is something oxymoronic, ungracious and even disreputable about a calculating kind of "generosity." And yet, it is very much part of the human condition. This may also be why unrequited love and kindness is the stuff of romantic mythology as well as high-toned religious injunctions and saintliness: because it isn't found very often among real people.

Reputation and *Richard III*

The New Testament is filled with exhortations about being "good," which is to say, behaving with loving altruism toward others. As a result, people scramble to associate themselves with these ideals, although actually practicing them is often a different matter.

In Shakespeare's *Richard III*, for example, there is a hilarious scene in which Richard—one of the most despicable, altogether unredeemable villains in literature—is trying to convince the citizens of London that he is ethical, altruistic, and trustworthy. So, he arranges to be seen reading the Bible, something he assuredly does not do when left to his own devices.

Richard III is an extreme case, but not unique. It is very much in everyone's interest to be considered reliable, someone who plays by the rules, who can be counted upon to pay his debts, not to cheat, and so forth. Like Richard III, politicians often parade their religious beliefs and presumed high moral sentiments in order to allay voters' fears that they might actually be venal and self-serving. Even in daily life, people strive to present themselves as moral and altruistic, whether via church attendance, conspicuous generosity, or—at minimum—maintaining a reputation for "family values." Concern with reputation is widespread among *Homo sapiens*, and when not based on fear of retaliation ("don't tread on me"), such reputation is usually designed to suggest that one is worthy of an altruistic investment, even if not genetically related.

Not surprisingly, mutually beneficial reciprocating exchanges are especially likely to develop in small, comparatively intimate social networks, within which everyone knows everyone, and where no one can afford to alienate a potential long-term reciprocator. At the same time, "con artists" are nearly always strangers, who prey anonymously on their victims and then move on.

But reciprocating systems aren't limited to situations of just two "players." There is, for example, a potentially important phenomenon, dubbed the "reputation effect" or sometimes, "indirect reciprocity." It may help provide a non-reciprocal avenue to altruism between non-relatives. When vampire bats share blood, or male baboons are called upon to repay a favor on demand, they are expected to reciprocate directly to the individual who earlier had assisted them. Otherwise, it is thought, the original altruist will not be reimbursed. But what if others are watching when someone does a good deed? These bystanders are then in a position to reward the altruist and also to punish those "ingrates" who receive benefits and then don't reciprocate. The existence of bystanders, then, provides both added opportunity as well as motivation to be considered an altruist, moved only by the "purest" (that is, most unselfish) desires. Certainly, it beats the ostracism that follows being seen as a selfish sonofabitch.[24]

This reputation effect may help explain why people sometimes behave altruistically when the opportunity for repayment is not immediately obvious. Consider donating blood, for example, or giving money to the March of Dimes, or the local National Public Radio station. Such generosity does not benefit one's relatives, and even reciprocal altruism would not seem to apply. After all, it is easy to "cheat," to receive the benefits of other people's donation, without incurring any of the cost. (Recall the "free rider" problem, and the question, "Who pays for the lighthouse?") And yet, people do give blood to strangers and money to causes they believe in. Maybe these are simply examples of the nonadaptive and incomprehensible vagaries of an unpredictable species. It is also possible, however, to identify a likely genetic logic at work.

First, note that most people do *not* donate in this manner; that is, impersonally, or in large amounts. That's why charities are notoriously under-funded, and require special tax breaks, as well as appeals to personal guilt. NPR is not swimming in money, nor is the Red Cross typically awash in donated blood. Moreover, people who are especially generous or devoted to charitable causes tend to be those who have had personal or family experience with the problem at hand. The Kennedy family, for example, has long supported mental health and retardation programs; it is not coincidental that one of the Kennedy sisters was mentally retarded. Betty Ford—herself a recovering alcoholic and victim of breast cancer—has been active on behalf of programs fighting breast cancer and alcoholism.

Even among those whose involvement seems wholly disinterested, there may nonetheless be a payoff in being *perceived* as an altruist. Not surprisingly, charities find that donations increase when people are given some sort of sticker or insignia that identifies them as donors. There is a benefit, it seems, in identifying oneself as generous, perhaps because others see the donor as a better, more worthy person, more deserving of their friendship, etc. Such diffuse paybacks are especially important in generating pressure for "moral" behavior. Often, people behave in a manner that appears to be unselfish in that there is no obvious, immediate genetic benefit, but in such cases, the altruist can generally count on receiving an indirect return, either from a grateful society, or from specific individuals who perceive the altruist as an especially meritorious person.[25]

Imagine that a house is on fire. A passer-by runs inside and rescues a toddler who had been trapped in her upstairs bedroom. The rescuer is a hero, and, if unrelated to the victim, might appear to invalidate the underlying biological selfishness of altruism, especially if there is no expectation or possibility of reciprocity. (And after all, what can a two-year-old do to repay her rescuer?) But similar acts of heroism may well have been adaptive in humanity's long evolutionary past when a person's regular associates were likely to be either relatives or related to others capable of reciprocating for the hero and/or his family.

Not only that, but even in the modern, impersonal, technological world, the hero who rescues a child from a burning building is likely to have his or her reputation enhanced as a result. And of course, social tradition and cultural norms strongly encourage such acts, at least in part because while many people dream of being heroic, selfless, and good, it doesn't happen very often, not only because the opportunity is rare, but probably because the inclination is equally uncommon.

Retaliation, Revenge, and Einstein's Lament

Earlier, we looked at the "dark side" of nepotism; namely, intolerance, violence, and possibly racism, directed at those who are different (genetically, physically, or culturally). Also garden variety selfishness, with its more luxuriant pathological forms. It is time now to identify a comparable dark side of reciprocal altruism, namely reciprocal retaliation and revenge. It is interesting that, so far, revenge has only been described for two species: human beings and our closest relatives, the chimpanzees. Earlier, we described the tendency of chimps to begrudge favors to individuals who have been ungenerous in the past. Now, we consider something more direct, more vigorous and—like it or not—more "human."

For chimps, the basic observation is as follows: lets say that individual 1 has been in the habit of intervening frequently against 2, helping a competitor get bananas, or matings. In this case, individual 2 is likely to be found returning the disfavor by intervening against 1 when the opportunity arises.[26] Moreover, if one chimp does not reciprocate, after another has gone out of its way to be helpful, then the other may retaliate. The result is a kind of inverted reciprocity, in which grudges can be as real as gratitude. Interestingly, the effect isn't all bad, especially if it serves to limit the amount of nastiness to which everyone is subjected. Thus, since chimpanzees "keep score" and will retaliate if picked on, they are probably less likely to be taken advantage of in the first place. In the course of keeping potential victims from being excessively victimized, grudge-keeping and the potential for retaliation may also serve to restrain the despotism of those on top. (On the other hand, it could also set the stage for the bullying of individuals who are unable or otherwise unlikely to retaliate.)

It is interesting to note that chimpanzees are actually rather egalitarian in their social organization. By contrast, the more rigidly hierarchical monkeys such as macaques or baboons do not participate in revenge, maybe because they cannot remember who-did-what-to whom, but more likely because their social networks are so clearly defined that it would be almost inconceivable for a subordinate to retaliate against a dominant. But among chimps, in which there isn't that much difference between the highest and the lowest in the pecking order, the opportunity exists for any high-ranking individual to be taken down a peg or two, if he (and usually it is a "he") goes too far.

Knowing that even a subordinate is likely to retaliate—if not strictly an eye for an eye and a tooth for a tooth, then perhaps a tooth for an eye—might restrain a dominant chimpanzee, or the closely related human being, from pressing his advantage or cheating excessively.

If we next add kinship groups to the social/biological equation, in which whole families may respond to other families with cordial reciprocity or begrudging animosity, new light is shed on how the Hatfields and McCoys, or Democrats and Republicans, or even whole nations, have come to exist.

Here in the bowels of retaliation, grudge-keeping, recompense and reward may also be found some rudimentary prerequisites for democracy, or at least, a social system in which each individual counts for something. It may also contribute not only to the origin of human society, but—as already suggested in a more positive context—also the human brain. Individuals in a complex, reciprocating society must have sufficient memory and awareness to calculate appropriate exchange relationships (recall our earlier discussion of fairness), and to respond accordingly. They must also have a marked capacity for behavioral inhibition, the ability to bide one's time and calculate longer-term payoffs.

Such calculations require no small amount of brainpower, or at least enough cognitive ability to keep good mental records of who has helped you and who has harmed you, who you owe, who owes you, and how much. If there were sufficient evolutionary rewards to successful book-keeping of this sort—as well as sufficient costs to blundering—then the unusually large brains of *Homo sapiens* may have been due, at least in part, to the payoff from making successful calculations about such payoffs. Indeed, the capacity to inhibit one's behavior and to keep multiple abstract concepts in mind without taking immediate action are some of the crucial functions of the frontal lobes, brain structures that are especially large and well developed in human beings. (It may be noteworthy that vampire bats, current undisputed champion of animal reciprocators, are also the brainiest bats known.) It is altogether possible, in short, that the difficult demands of reciprocity have contributed mightily to making *Homo sapiens* so sapient.

Successful kin selection does not usually require deep and devious calculations, whether conscious or unconscious. The fact that the recipient is related and able to benefit from the altruism is generally sufficient. But reciprocity is different: the genetic make-up of the recipient will not be enough by itself, because from the "altruist's" perspective the crucial payoff comes not from benefiting someone else but only when (and if) there is a payback. It therefore helps to be able to calculate contingencies. In short, to be smart.

Thus, even though natural selection appears to act at the level of the individual—and more precisely, that of the gene—selection for a large, calculating human brain, with the capacity to keep images in "mind" while postponing action, may have provided *Homo sapiens* with the chance to act however we choose, even including "for the good of the group" if that is desired. Thus, natural selection has probably favored the evolution of mental equipment that permits complex, considered, ethical behavior, including, on occasion, selflessness—all in the service of "selfish" ends! The result may be an interesting example of how human beings have the capacity to transcend their biological heritage, by using the very hardware that natural selection has bequeathed to us.

Certainly, there is no shortage of social problems to which our species can apply its impressive brain. Among the difficulties that individuals, families, and whole societies face is not only when to be nice or nasty, but how nice or nasty to be, and for how long. When, for example, is it appropriate to forgive and forget, to send a dysfunctional relative to an expensive treatment center, or to start a Marshall Plan? When should you cut off economic relations? Diplomatic relations? Christmas visits? How severely should a perpetrator be punished? Should murderers get the death sentence? How about child molesters? What if a guilty individual is of diminished competence? What about extenuating circumstances? When—if ever—is it appropriate to punch someone in the nose, to make friends, to cooperate, ostracize someone, or go to war? More generally, what obligations do people have toward each other? And more generally yet, what about obligations toward the rest of the living world? Toward posterity? The planet? God?

These dilemmas—largely compounded of kin selection mixed with reciprocity—are among the hallmarks of the human condition. And they are probably what Albert Einstein had in mind when he commented that politics—to which it seems reasonable to add interpersonal relations generally—is far more difficult than physics.

Notes

1. Stephen T. Emlen. 1981 Altruism, kinship, and reciprocity in the white-fronted bee-eater. In Natural Selection and Social Behavior, R. D. Alexander and D. W. Tinkle, eds. Chiron Press: New York.
2. Gerald S. Wilkinson. 1990. Food sharing in vampire bats. Scientific American, February, 76-82.
3. L. K. DeNault and D. A. McFarlane. 1995. Reciprocal altruism between male vampire bats, *Desmodus rotundus*. Animal Behaviour 49: 855-856.
4. Craig Packer. 1977. Reciprocal altruism in Papio anubis. Nature 265: 441-443.
5. R. Noe. 1992. Alliance formation among male baboons: Shopping for profitable partners. In A. Harcourt and F. B. M. de Waal, eds., Coalitions and Alliances in Humans and Other Animals. Oxford University Press: New York.
6. P. C. Lee. 1987. Allomothering among African elephants. Animal Behaviour 35: 278-291; Owens, D. D. and M. J. Owens. 1984. Helping behaviour in brown hyenas. Nature 308: 843-845; J. P. Rood. 1978. Dwarf mongoose helpers at the den. Zeitschrift fur Tierpsychologie 48: 277-287.
7. Tavolga, M. C. and F. Essapian. 1957. The behavior of the bottlenose dolphin *Tursiops truncatus*: Mating, pregnancy, parturition, and mother-infant behavior. Zoologica 42: 11-31; M. L. Reidman. 1982. The evolution of alloparental care and adoption in mammals and birds. Quarterly Review of Biology 57: 405-435.
8. M. C. Caldwell and D. K. Caldwell. 1966. Epimeletic (caregiving) behavior in Cetacea. In Whales, Dolphins and Porpoises. K. S. Norris, ed. University of California Press: Berkeley.
9. David Hume. 1777 (1975) An Enquiry Concerning the Principles of Morals. Clarendon: Oxford
10. John Archer. 1997. Why do people love their pets? Evolution and Human Behavior 18: 237-259.
11. G. Teleki. 1973. The Predatory Behavior of Wild Chimpanzees. Bucknell University Press: Lewisburg, PA.
12. C. B. Stanford, J. Wallis, E. Mpongo, and J. Goodall. 1994. Hunting decisions in wild chimpanzees. Behaviour 131: 1-18.
13. A. Zahavi. 1995. Altruism as a handicap—the limitations of kin selection and reciprocity. Journal of Avian Biology 26: 1-3.
14. David P. Barash and Judith Eve Lipton. 1997. Making Sense of Sex. Island Press: Washington, DC.
15. John H. Beckstrom. 1987. The use of legal opinions to test sociobiological theory: Contract law regarding reciprocal relationships in a household. Ethology and Sociobiology 8: 221-247.
16. R. M. Seyfarth and D. L. Cheney. 1984. Grooming, alliances and reciprocal altruism in vervet monkeys. Nature 308: 541-543; D. L. Cheney and R. M. Seyfarth. 1990. How Monkeys See the World. University of Chicago Press: Chicago.
17. Frans De Waal. 1996. Good Natured. Harvard University Press: Cambridge, MA, 153.
18. R. Axelrod. 1984. The Evolution of Cooperation. Basic Books: New York.
19. Frans De Waal. Good Natured.
20. Thanissaro Bhikkhu. 1996. The economy of gifts. Tricycle, the Buddhist Review. 6: 56-58.
21. Quoted in Bhikku.

22. R. L. Trivers. 1981. Sociobiology and Politics, in E. White, ed. Sociobiology and Human Politics. University of Toronto Press: Toronto.
23. Richard D. Alexander. 1979. Darwinism and Human Affairs. University of Washington Press: Seattle.
24. Robert Boyd and Peter J. Richerson. 1992. Punishment allows the evolution of cooperation (or anything else) in sizable groups. Ethology and Sociobiology 13: 171-195.
25. Richard D. Alexander. 1987. The Biology of Moral Systems. Aldine De Gruyter: Hawthorne, NY.
26. Frans B. M. de Waal, and L. M. Luttrell. 1988. Mechanisms of social reciprocity in three primate species: symmetrical relationship characteristics or cognition? Ethology and Sociobiology 9: 101-118.

5

Parenting, Adoption, and Step-Parenting

Each of us is a genetic slingshot, a catapult that shoots genes into the future. This is what living things are about, and from a strictly evolutionary perspective, it is *all* they are about, the only reason they exist. This helps explain why human beings are not entirely encompassed by evolutionary insights; we are unique among living things in being able to say no to various biological imperatives, breeding not the least. The fact that we are *capable* of saying no, however, does not mean that we typically do so, or that it always feels good.

Living things generally have not been shy when it comes to availing themselves of the evolutionary payoff that comes with gene-slinging, that is, reproduction. Every salamander, every seagull, every human being alive today is descended from a long unbroken line of breeders. Looking back at your own ancestral lineage, through parents, grandparents, great-grandparents, etc., not one of them ever missed a beat! Small wonder that living things, by and large, are committed to reproducing. (Those that were otherwise committed generally didn't leave descendants to carry on in their genetic footsteps.) Small wonder, as well, that for all the variety of social, economic, and political systems devised by human beings, there is not a single society in which people routinely give up their reproduction to someone else. People indulge in all sorts of specialization and division of labor, but propelling genes into the future is something nearly everyone chooses to do for him or herself.

Indeed, the clear-cut biological significance of reproduction is why altruism—whether kin selected or reciprocal—is so interesting: because altruism, when we first encounter it, appears to go against this

139

most basic principle of the living world. Being a parent, by contrast, is so obvious, so appropriate, so biologically *de rigeur* that up until recently it has largely escaped the scrutiny it deserves.

The obviousness of reproduction and parenting derives from one simple fact: it is the most straightforward way for individuals to enhance their evolutionary success, or—more to the point—for genes to project copies of themselves into the future. If you want something done right, goes the saying, do it yourself. Nowhere is this more true than when it comes to gene-throwing. Whenever offspring are being produced, fed, trained, kept warm or cool, wet or dry, protected from enemies or introduced to friends, something fundamental is going on: genes are nurturing copies of themselves.

"He that hath wife and children," wrote the sixteenth-century English philosopher Francis Bacon, "hath given hostages to fortune, for they are impediments to great enterprise, either of virtue or mischief." Bacon, one of the great architects of modern science, lived too early to understand this important finding of evolution: Children are not impediments but passports to the most pressing enterprise—indeed, the only enterprise—of life itself, whereby genes promote their own success. In this respect, all living things are hostages, not to fortune but to natural selection.

"We had lots of kids, and trouble and pain," goes the folk-song, *Kisses Sweeter Than Wine,* "but Oh Lord, we'd do it again." Why would they do "it" again, given that having kids involves so much trouble and pain? And why did they do it the first time? Hint: It isn't simply because of those oh-so-sweet kisses. Rather, natural selection has contrived to make romance and love and sex and kisses "sweet" in the first place because it is a way to encourage people to take part in the whole endeavor.

Try asking spiders of the species *Stegodyphus mimosarum,* found in Africa. They carry parental self-sacrifice a bit far.[1] Comfortably housed within silken nest chambers woven by the mother just for this purpose, baby spiderlings perch on her abdomen and cheerfully fill their own bellies with their mother's "flesh," eventually killing her in the process.[2] Talk about trouble and pain!

Since the mother *Stegodyphus mimosarum* cannot answer for herself, let's do it for her. She reproduces, and then allows—probably even encourages—her offspring to kill and eat her because they are so irresistibly cute. Or because it feels so wonderfully good, maybe like having an itch scratched. Or maybe because there is simply no alternative, like the mammalian need to breathe, or circulate blood. Whatever the immediate mechanism, the end result is that her spiderlings are sent off into the world with a full stomach at the mother's expense, somewhat more extreme than the suburban parent making sure Junior goes off to school with a freshly made peanut butter and jelly sandwich—but at the most basic biological level, not altogether different.

Any time a parent reproduces sexually, it creates small replicas of itself, not precise duplicates to be sure, but half-copies. Not half-*size* copies, but half-copies in a more profound sense, a genetic sense. From the viewpoint of any given gene within a parent, there is a 50 percent chance that its identical replica resides inside each child. This is why living things reproduce, African spider or American suburbanite, as well as everything else in between. It is also the primary reason for love, including love of parents for children as well as of parents for each other. And it goes a long way toward telling us why males and females differ in their style of being parents, as well as why step-parenting is so often a conflicted and difficult business.

Who Takes Care of the Kids?

Among mammals, the answer is simple: the mother. Fathers sometimes help, but only in a tiny minority of species. Even then, the most they do is *help*. There is not a single kind of mammal in which males do more parenting than females. This, in turn, could be due to the fact that among mammals, females are uniquely equipped to nourish their offspring, via their milk. Could that be why male mammals are much less paternal than mothers are maternal? Because they just can't contribute as much?

Most likely, this is only a partial answer, and a misleading one at that. It is worth asking, for example, *why* females are so uniquely adapted to lactate. After all, males even have rudimentary nipples. After their mates have gone through the demanding, stressful experience of pregnancy and childbirth, why don't fathers take over, and do the nursing? It would seem only fair. Wouldn't it also be adaptive?

Adaptive, however, for whom?

It would indeed be "only fair" for fathers to nurse, but it would not be in *their* best interest. Because lactation is expensive, only a guarantee of shared genes can justify the enormous investment that it entails.

The lactating individual must be nothing less than certain that the offspring benefiting from such heavy investment is genetically related to the investor, the one providing the energy capital. (In evolutionary terms, genes within a lactating individual must know that in committing the body to making milk with all this entails, they are helping copies of themselves.)

The key appears to be confidence of genetic relatedness: once again, not necessarily conscious confidence, just a high probability that the underlying connection is there. A look at parental behavior in other species helps confirm that shared genes is the key to parental behavior.

All living things reproduce, but many are not especially parental. Oysters, for instance, just squirt their eggs and sperm into the sea. Most insects, amphibians, and reptiles, after engaging in elaborate courtship rituals, simply deposit their fertilized eggs, then go on with their lives—or deaths—leaving the offspring to find their own fate, unaided. When parental care does occur, confidence of relatedness seems to be the most important factor that determines who does the job. Significantly, when fertilization takes place outside the female's body, as in many fish, males and females are about equally likely to care for the young. In such cases, the two sexes are pretty much in the same boat when it comes to knowing whose genes are tucked away in the developing embryos, and who may have been cuckolded by a passing stranger.

It is when fertilization is internal that parental care becomes consistently asymmetric. And not surprisingly, when this asymmetry occurs—when females experience a kind of genetic advantage over males, a unique certainty of relatedness—females bear the brunt of the parenting. When their eggs are fertilized inside their bodies, females "know" that their offspring are really their own: "Mommy's babies," goes the old saying, "daddy's maybes." When a bird lays an egg, or a mammal gives birth, the presumptive mothers are guaranteed to be mothers in fact; there are no known cases of females sneaking their eggs into the reproductive tract of other, unsuspecting females, who are then duped into thinking themselves related to offspring which are not really their own.

The situation is exactly the opposite for males. When fertilization takes place within the female's reproductive tract, males can only hope to be fathers, trusting to their mates' fidelity, their own watchfulness, ability to intimidate would-be Lotharios, or just plain good luck. Moreover, even among those relatively few species thought to be monogamous, it nearly always happens—at least on occasion—that females bear young which have not been fathered by their mates.

Researchers studying chimpanzees of the Tai Forest, in West Africa's Ivory Coast, collected DNA samples from all fifty-two members of a group. When they conducted paternity tests on thirteen infants, it was revealed that only seven had been fathered by males from their particular group. Female chimps had evidently been sneaking off and mating with males from other groups.[3] It is, incidentally, a mystery at present why some female chimps consider it so important to mate with comparative strangers, especially since they do so at some risk, if caught in the act by the cuckolded males. One possibility is that by mating with neighboring males, they are buying tolerance from these same males if and when her offspring encounter their potential fathers.

In any event, whereas male mammals can be cuckolded—induced to spend time and effort rearing someone else's young—it is unheard of for females to be comparably deceived. At the same time, there is a disadvantage associated with the genetic confidence of females: They are the ones who get stuck with the child-care.

They are not necessarily condemned to single-parenthood, however. Among many birds—especially the most abundant order, the Passerines or perching birds, which includes the common songsters such as robins, sparrows and warblers—males and females cooperate in rearing the young, dividing their parenting chores pretty much equally.

Revealingly, biparental care typically occurs among species in which the young grow at an especially rapid rate, making phenomenal demands on the hard-working parents, who may have to bring insects to the ravenous nestlings at a rate of one trip every twenty seconds or so. Under these conditions, assistance from the feathered father may well be mandatory: a single-parenting female warbler has virtually no chance of rearing any offspring at all. Male warblers, therefore, have little choice but to pitch in. But it surely is no coincidence that in such cases, the pair bond is generally monogamous, so that the hard-working male has a high confidence that he is the father—although to be sure, not as high as the female's confidence that she is the mother.

A revealing twist is provided by some provocative studies conducted on a small European song-bird, the pied flycatcher. In this species, male and female typically cooperate bringing food to the hungry nestlings. Not surprisingly, the male pied flycatcher is relatively confident that the nestlings he helps to rear are, in fact, his. In a field experiment, a trio of Scandinavian researchers removed the males from twenty mated female pied flycatchers after they had been fertilized, thereby rendering them potential single parents. Within two days of their "widowhood," six of these females were seen soliciting copulations from new males, and in a few cases, these males were seen bringing food to the nestlings that eventually appeared, even though they could not have been the fathers.

It is important to note that under normal conditions, female pied flycatchers become utterly indifferent to sex, if not downright hostile, once their eggs have been fertilized. Thus, the newly developed sexual avidity on the part of the researchers' female subjects was almost certainly a ploy on their part (although probably not a conscious one) to fool a male into mistakenly acting as though their nestlings are also his.[4]

It is interesting to note that those birds in which the males are less likely to assist with child-care, such as members of the pheasant or grouse family, are also likely to be non-monogamous. For mammals, the pattern is similar, but if anything, more extreme. Thus, fatherly assistance is very rare among mammals, consistent with the generalization that young mammals grow more slowly, as a percentage of their body weight, than young birds. Young mammals therefore require proportionately less parental—specifically, paternal—aid than do young birds. The fact that male mammals are by and large less paternal than male birds is also in line with the fact that female birds— no matter how motherly—don't lactate whereas even the least maternal of mammals do. Furthermore, in those unusual cases' in which father mammals do pitch in to provide significant care for their young, they always have good reason for confidence that those young are their own. Not surprisingly, these two rarities—monogamy and paternal care—coincide: Paternal care is pretty much limited to monogamous species.

It is a good guess that by and large, fathers also find it easier to let go than do mothers. In "Walking Away," the poet C. Day Lewis gives a predictably male view when he describes his eldest son going away to school, and the poet's awareness of "How selfhood begins with a walking away/ And love is proved in the letting go."[5]

As to maternal care, it—like paternal care—is not simply broadcast indiscriminately, available to anyone who shows up. During biology's Dark Ages, before gene-centered thinking gained prevalence and when it was widely assumed that animals did what they did for the good of the species, there was little expectation that parental solicitude would be limited to genetic offspring. After all, if the species benefit from all-purpose benevolence, then parents should broadcast benevolence to any juvenile who could use it, rather than favoring their own offspring.

An interesting example was the Mexican free-tailed bat. Although each female produces only one young at a time, these animals congregate in immense gatherings, hundreds of thousands and more in a single cave. While the adults are out cruising for insects, the young crowd together in creches, with population densities as high as two pups per square inch! Utter chaos appears to reign, especially when females return from foraging and the young swarm all over them. Consistent with good-of-the-species thinking, instead of its good-of-the-gene alternative, batologists had long thought that the young were fed indiscriminately and rather communistically: from each according to her ability, to each according to his (or her) need. Looking at the melee, it is difficult even for a modern biologist to imagine how parent-offspring pairs are ever sorted out.

But when Gary F. McCracken of the University of Tennessee at Knoxville studied these nurseries, he found that female Mexican free-tailed bats were not at all free when it came to dispersing milk. Mothers recognized their own young 83 percent of the time, apparently by sound and smell. Whenever a female suckled young not her own, it was apparently a result of error (on her part) and/or milk-stealing (by the little batling), not design by the mother or by natural selection.[6]

Even in those rare cases when parents do not appear to discriminate in favor of their own offspring, there is an underlying method to their seeming evolutionary madness. For example, among lesser snow geese, adults readily accept extra goslings, appearing to be genuinely impartial when it comes to doling out their parental solicitude. The reason seems to be that large goose "families" are socially dominant over small families; as a result, the larger the family the more food everyone gets. And so, it pays Mother Goose—and Father Goose as well—to be especially generous to stray goslings, not because of disinterested altruism, but because their own goslings do better as a result.[7] It is not clear whether adult geese are unable to discriminate their goslings from unrelated strangers, or whether they simply refuse to discriminate against the latter because such benevolence is beneficial to their own offspring.

In any event, the situation of lesser snow geese is lesser indeed: dramatic parent-offspring discrimination is overwhelmingly the norm among vertebrates, and is a specific case of the more general phenomenon of kin recognition, or genes recognizing copies of themselves traveling incognito, in other bodies. It also illuminates the surprising capacity of human beings to engage in a special kind of parenting, one that seems to contradict evolutionary wisdom: adoption.

Recognizing One's Offspring, and Adoption

First, an important rule: natural selection doesn't do more than is needed. Thus, it strongly promotes mechanisms that enable parents to recognize their offspring, but only if there is a threat that otherwise, parental care would be misdirected. (Analogously, selection has outfitted camels with the ability to go for a long time without water, but not all animals are so endowed: only those living in deserts.) When their way of living makes it unlikely that parental care will be wasted on someone else's offspring, animals are not equipped with special recognition mechanisms. This evolutionary minimalism occurs because "selection pressure" is needed to create something out of nothing. No pressure, no adaptation.

Parent-offspring recognition should therefore be acutely developed when mix-ups are likely, and absent when they aren't. For a good test case, Mike Beecher of the University of Washington turned to a pair of bird species, the bank swallow and the rough-winged swallow. Bank swallows nest in burrows dug in clay banks and are colonial, with many nesting pairs closely associated. Hence, this species runs the risk that parental care might be misdirected to someone else's young. Rough-winged swallows, on the other hand, are essentially solitary, each pair maintaining a nest that is isolated from others of their species. So there is very little chance that a rough-winged swallow will accidentally proffer food to nestlings other than its own.

Beecher found that the vocalizations of young bank swallows (the species vulnerable to mix-ups) are more distinctive than those of roughwings. Having a unique "vocal fingerprint" makes it easier for bank swallow parents to learn the distinctive vocal traits of their offspring. At large bank swallow colonies, nest entrances are very near each other. As a result, youngsters often land at the wrong nest. When they do, the adults shoo them away, reserving food and protection for their own offspring. By contrast, rough-winged swallows are undiscriminating. They can afford to be, because under normal conditions, they run no risk of being parasitized by strangers. And so, rough-winged swallows can be fooled by an experimenter, induced to accept foreign nestlings introduced into their nest, something that bank swallows never do. Rough-wing parents will even accept bank swallow babies; bank swallow parents won't accept rough-winged babies.[8]

Does this have anything to do with human beings? Indeed it does. In every known human society, mothers are more maternal than fathers are paternal. And of course, for humans just as for all mammals, mothers—and not fathers—can be confident that they are related to their offspring. (Translating into gene-centered language: genes present in mothers can induce whole-hearted devotion to offspring, because maternal care such circumstances could not have been misdirected. By contrast, genes present in fathers would be selected to be more circumspect, since paternal care could be "wasted" on another man's children.)

Evolutionary psychologists have even found that mothers are especially likely to point out that their newborn infant resembles the father: "Look, dear, he has your chin," etc. In the musical, *Oklahoma!*, the fiancé of Ado Annie—the "girl who cain't say No"—sings about their future child, warning his betrothed: "He better look a lot like me." There is no need, and hence, little incentive, to point out a resemblance between an infant and the mother. It would make no sense for Annie to demand that their future child had better resemble her!

It is also noteworthy that the tendency to talk up the physical resemblance of father and child, initially discovered among Canadian parents, was found among the mother's family, but not the father's; once again, mothers and their relatives have an interest in persuading the marriage father that he is indeed the genetic father, thereby inducing him to behave paternally, or at least, more than he might otherwise.[9] The same pattern was subsequently confirmed among Mexican families.[10]

There is even some reason to think that young children really *do* resemble their fathers more than their mothers! A research study presented photographs of young children along with, in each case, three different possible mothers and three different fathers. Based on chance alone, unknowing observers should have paired each child with the correct parent 1/3 of the time. This was the case when the children in question were ten-year-olds and twenty-year-olds. But in the case of one-year-olds, observers correctly paired children with their fathers nearly 1/2 the time, although a similar effect was not found with mothers.[11] These findings are only preliminary, but it would not be altogether surprising if natural selection, acting on the physical traits of children, has produced a tendency for them to show a special physical resemblance to their birth fathers. This could readily happen if through evolutionary time, individuals with such resemblances have been more likely to receive paternal care, and thus, to be more successful than others whose physical appearance was more ambiguous, and who therefore evoked more indifference—if not outright hostility—from their fathers.

In another fascinating series of studies, researchers took a different tack in examining the connection between confidence of genetic relatedness and the tendency to provide assistance. Once again, the starting point is the fact that mothers have much higher confidence of relatedness than do fathers. Looking across generations, then, at the chain of genetic confidence linking any two individuals, fathers can be expected to introduce more uncertainty than mothers. Unlike a mother, each father is, in a sense, a genetic weak link. Every child has four genetic grandparents: mother's mother (MoMo), mother's father (MoFa), father's mother (FaMo) and father's father (FaFa). In the case of MoMo, there is utter confidence of genetic continuity from grandmother to grandchild, since the maternal link is guaranteed. For MoFa and FaMo, there is one link of paternal uncertainty. And for FaFa, genetic confidence is lowest, since there are two such weak links. The following predictions can therefore be made, examining relationships between grandparents and their grandchildren: MoMos should be more grandparental than MoFas or FaMos, whereas FaFas should be the least grandparental of all.

When 120 American undergraduates were surveyed about their relationships with their grandparents, they consistently ranked closeness, amount of time spent together, and grandparents' financial contributions in precisely this order: MoMos were the most involved and the closest, FaFas were the least, and MoFas and FaMos were intermediate.[12] Even many evolutionary psychologists have been surprised by the precision of this finding, a remarkable confirmation of how biology—unbidden and until recently, unknown—subtly influences our personal lives. After these results were presented, the same pattern was independently confirmed in a study of German students. suggesting that it is not a cultural artifact but a result of biological prodding.[13]

The burden of these and many other studies is consistent: people, like other animals, behave parentally—and even grandparentally—in ways that are predicted by genetic relatedness, and by the confidence, and lack of confidence, that such relatedness is real. (It may be possible for ideas—even important and beloved ones—to be confirmed *too* closely, simply because we all know that biology alone does not and cannot tell the whole story. Thus, it was with something approaching relief that sociobiologists learned of a study that failed to confirm female-biased "discriminative grandparental solicitude" in rural Greece.[14] After all, even though both biology and culture are crucial in all cases, either can override the other. Imagine that a society suddenly conferred great rewards on people with an eye in the middle of their forehead; despite the most enthusiastic cultural prodding, it is unlikely that an increased number of little cyclopses would accordingly be born. Similarly, it is not surprising—and even reassuring—that despite the likely whisperings of biology, in a rigidly patrilateral cultural such as rural Greece, male-biased grandparental involvement remains the norm.)

What about adoption? After all, to adopt is to expend time and resources on behalf of someone *unrelated* to the adopter. Accordingly, it would appear to be an evolutionary blunder, comparable to genuine altruism (that is, beneficence toward another without the compensations of either kin selection or reciprocity). And yet, human beings are quite capable of adopting, often struggling against heavy odds and bureaucratic red tape to do so. Adopted children, moreover, are generally well cared for, and about as successful as biological children.

First, let's point out the obvious: Adoption, overwhelmingly, is *not* most people's first choice. When it comes to children, the vast majority prefer to make their own, given the opportunity. If this is not an option, then people may be inclined to satisfy their desire to be a parent—a desire that is almost certainly a highly adaptive legacy of evolution—by parenting someone else's kids. Bear in mind that for perhaps 99.9 percent of its evolutionary past, the mammal species known as *Homo sapiens* lived in small hunter-gatherer bands that almost certainly numbered fewer than 100. Within such groups, most individuals were related. As a result, anyone who adopted a child was likely to be caring for a genetic relative (hence, kin selection). Even individuals who cared for an unrelated child may well have positioned themselves to receive a return benefit from the child's genetic relatives as well as becoming a possible recipient of reciprocal altruism or indirect reciprocity.

Furthermore, when it comes to the question of offspring recognition, people are bank swallows rather than rough-wings. After a woman "has" a child there is simply no question whether the baby is hers. Switching infants at birth may take place in Gilbert and Sullivan operettas or—very rarely—in a modern, crowded metropolitan hospital, but not among a small band of early hominids living on the Pleistocene savannah. Lacking the threat of misidentifying our babies, our ancestors also lacked an automatic, lock-and-key recognition mechanism. (Recall the minimalism of evolutionary adaptations.) As a result, we have a wonderfully "open program" when it comes to identifying children as our own. We rely on continued interaction to learn the identity of our close relatives, just as we can be open, even eager, to love and care for children, regardless of whose genes they carry.

Step-Parenting: Animals

Step-parenting, on the other hand, is a darker and more troubled phenomenon. It is similar to adoption, in that nongenetic "parents" end up taking care of someone else's children. But step-parenting is crucially different. Whereas adoption involves a specific commitment to the adopted child, step-parenting typically comes about as a side-effect of two adults' commitment *toward each other*. Step-children, if any, are thrown in as an unavoidable—and, if the truth be acknowledged, often unwanted—part of the deal. Adoption generally takes place when the child is an infant, thereby enhancing the prospects that adopting parents can "fool themselves" into responding as though the adopted child is genetically their own. On the other hand, it usually isn't until they are older that step-children enter the step-parent's life, which is a further obstacle to parental love.

Turning to animals, we find that once again they illuminate the human condition. The equivalent of adoption and step-parenting does occur, although not frequently. When animals associate "parentally" with young unrelated to themselves, the results are sobering, and consistent with the interests of the adults' genes.

There is a fresh-water fish known as the fathead minnow. Like many fish practicing external fertilization, male fathead minnows care for the eggs, which are glued by the female to the underside of rocks and then left to the male's paternal attention. Often, fathead minnows suffer from a shortage of good nest-sites, so males compete for reproductive real estate, sometimes evicting one who is already tending a batch of eggs. When this happens, the new male "adopts" the eggs he has taken over, defending them from egg-predators and coating them with an anti-fungal secretion, even though they were fertilized by the previous occupant. Why is he so nice to someone else's offspring? Why not eat them instead?

It turns out that even in this case of seeming altruism and nonparental benevolence, selfishness rears its evolutionary head: Female fathead minnows are more likely to lay eggs with a male who is already tending a batch. Therefore, by adopting his predecessor's embryonic youngsters, the step-parent male is actually being genetically selfish, increasing his chance of becoming a father in his own right, perhaps by advertising what a good father he can be. Almost certainly, his behavior toward eggs in a recently acquired nest is *not* motivated by general altruism, or a disinterested generosity, but by the selfish prospect of his own genetic gain.

This is strongly suggested by the observation that fathead males are not fastidious about nibbling some of their predecessor's progeny. One day after a takeover, more than 50 percent of the eggs are mysteriously gone (by contrast, fewer than 15 percent of the eggs disappear during a comparable time period when the genetic father is left in charge of his own brood). Fathead minnow males leave just enough eggs to serve as bait for a passing female, not as testimony to their step-parental benevolence.[15]

A similar theme can be discerned among other animals. In one experiment, for example, mountain bluebirds were provided with nestboxes designed to be especially attractive to them. Bluebird pairs quickly moved in and started families, after which the males were removed, leaving the females single parents. They did not remain single for long, however. In what passes for bluebird society, they were "wealthy widows" since good nests are hard to find and each of these females owned a valuable piece of property. They and their dependent offspring were soon joined by fortune-hunting males. Significantly, these new arrivals—step-fathers of the nestlings—did not participate in feeding the youngsters, and only one in twenty-five gave alarm calls in response to possible predators.[16] When it comes to caring for their own biological offspring, that is, repositories of their own genes, bluebird males are much more attentive.

As with fathead minnows, male bluebirds in this case behaved selfishly. They affiliated with the single-parent females in order to profit, eventually, from their valued nest-sites, not out of love for their (that is, the females') existing offspring.

Among a common European songbird known as the dunnock, a family may consist of a female and two unrelated males, one of which does most of the copulating. If the second male does not copulate at all, then he doesn't aid in rearing the young; if he does, then he does.[17] The possibility even exists that female dunnocks permit occasional copulations by the second male in order to induce him to help out with the children, some of which might be his.

Something very much like step-parenting takes place among certain nonhuman primates. In these cases, the imprint of kin selection is usually to be found, and when it isn't, plain old-fashioned selfishness is far more likely than genuine, disinterested altruism. A review of the situation among primates concluded that those most likely to "adopt" are, in descending order of frequency: older siblings, close relatives other than siblings, adult males who may be the father, and—a distant fourth—unrelated adult females.[18] In one species—the hamadryas baboon, found in arid regions of Ethiopia—young males have been known to adopt young, unrelated females. Such behavior is closer to abduction than adoption, however, since the male's goal is hardly altruistic: He will eventually mate with his wards.[19]

Step-Parenting: People

With the backing of biology, and the consistent example of numerous animal species, is anyone surprised to find that among human beings, too, step-parenting is a conflicted affair, typically less benevolent than genetic parenting? This is not to say that step-parents cannot be loving, or that genetic parents are not sometimes abusive, only that biological pressures tend to be consistent, and the results, difficult at best.

"All happy families are happy in the same way," wrote Leo Tolstoy, in the famous opening sentence of *Anna Karenina*. "Every unhappy family is unhappy in its own way." As to happy families, there is room for debate, but when it comes to unhappiness, Tolstoy was certainly on to something: people have devised—or blundered into—innumerable ways of being unhappy. Even this diversity, however, resolves itself into some recognizable patterns. And one of the most common (and biologically predictable) concerns step-parenting.

Anthropologist Mark Flinn of the University of Missouri examined the family lives of residents in a small, rural village in the Caribbean nation of Trinidad. He was especially interested in the phenomenon of step-fathers, so he examined households with and without them. When both genetic and step-children were present, Flinn compared fathers' behavior toward the two. The results were strikingly in accord with biological expectations. For example, Flinn noted that when genetic and step-children live in the same household, fathers interact more frequently with their genetic offspring than with step-offspring. The only exception concerned aggressive, conflict-laden behavior (arguments, threats, punishments, etc.); here, the pattern switched, with step-fathers engaging their step-children more than their genetic children.

This finding cannot automatically be attributed to biology, of course. For example, there is a large body of so-called "attachment theory," which maintains that people are more affiliative toward others as a simple result of spending time with them in the past. Not absence, but interaction makes the heart grow fonder. It is likely that genetic fathers have been with their children since birth, whereas step-fathers presumably joined the household somewhat later (perhaps after the genetic father died, or went away). Maybe this explains the difference. Unfortunately for this appealing, common-sense notion, however, Flinn found that although it applied to genetic fathers—time living together correlated with more positive interactions and fewer negative ones— when it came to step-fathers, more time together actually correlated with *lower* rates of benevolent interaction, and *higher* rates of negative encounters.

Flinn's other findings also point in the same biological direction. Of ninety-seven children who did not have a step-father, ninety-four lived with their mother (97 percent). Of fifty-eight children who had a step-father, only thirty-six lived with their mother (62 percent); the rest had moved in with grandparents or aunts/uncles. Thus, the presence of step-fathers seems to drive step-children away.

Interestingly, the immediate presence of mothers reduced the hostility of interactions between step-father and step-child. With mothers present, 1.8 percent of these interactions were antagonistic; when mothers weren't nearby, the percentage of antagonistic interactions between step-fathers and step-children shot up to 9.4 percent, a fivefold increase. By contrast, interactions between fathers and their genetic children were not affected by whether the mothers were present. This is presumably because biological fathers are more likely to behave benevolently in any event, whereas in the case of step-fathers, antagonisms are leavened somewhat when the child's mother is nearby.

Finally, anthropologist Flinn found that Trinidadian step-children show several other signs of being discriminated against: they don't inherit land, they are more likely than genetic children to emigrate from their natal village, and finally, if they stay, step-children end up producing fewer of their own offspring than do genetic children. Step-children, in short, generally do worse—socially and biologically—than their biological counterparts.[20]

The Trinidad study was designed to test the evolutionary principle that genetic relatedness would be associated with closer, more affiliative father-child behavior, and that the absence of shared genes would correlate with greater social distance in the case of step-father and step-child. There is no reason to think that one particular village is unique on the island of Trinidad, that Trinidad is unique in the Caribbean with respect to step-fathering generally, or—more to the point—among most people around the world.

Here is some highly charged dialog between Yakov Bok, falsely imprisoned in a Russian jail, and his estranged wife, from Bernard Malamud's novel, *The Fixer*:

> "I've come to say I've given birth to a child."
> "So what do you want from me? . . . "
> " . . . it might make things easier if you wouldn't mind saying you are my son's father . . . "
> "Who's the father . . . ?"
> " . . . He came, he went, I forgot him. . . . Whoever acts the father is the father . . . "

From Trinidad to Russia, the truth is just the opposite: Whoever *is* the father is more likely to act the father. (The same applies, by the way, to mothers, but as we have seen, mothers are less likely to be deceived.)

Parenting can be difficult and stressful in the best of circumstances. But it is a truism that devoted biparental families are more likely when husband and wife are the biological parents of the children in question (all happy families are happy in the same way?). As with other animals, the key seems to be confidence of relatedness.

To be sure, there are benevolent step-parents. But the world's folk wisdom consistently fingers step-parents as potential trouble-makers. Folk tales, from Cinderella to Snow White, identify the "evil step-mother" or step-father, indifferent at best to the fate of his or her step-children, often in dramatic contrast to the love and solicitude lavished on genetic offspring.

According to the folk-tale, Cinderella was the abused step-child of an uncaring step-mother, who had her own two unpleasant daughters. Cinderella's mother had died, whereupon her father remarried, choosing as spouse someone who seemed to dote on Cinderella—until the father also died, whereupon Cinderella's step-mother revealed her true colors.

As to the universality of step-parenting stories, consult the immense, six-volume compendium, *Motif-Index of Folk Literature*, which provides a thorough compilation of the world's mythologies and folk tales.[21] There, doubters will encounter tale after tale featuring evil step-mothers, but will search nearly in vain for comparable accounts in which step-mothers are benevolent or well-intending. As to step-fathers, the *Motif-Index* identifies two categories: "cruel stepfathers" and "lustful stepfathers."

What are we to make of this? It is fruitless to deny the ubiquity of the evil step-parent image. But maybe social expectations are to blame. Maybe the *myth* of the malevolent step-parent is the cause of the problem. Maybe it is because step-parents are widely seen to be so difficult that they in fact are difficult. But this begs the question: *Why* are step-parents so widely considered in such a negative light? Most likely, step-parents are perceived as potentially dangerous and liable to treat their step-children badly because—all over the world—they are. (This is not to say that all, or even most, step-parents are cruel, lustful, nasty, neglectful or murderous. Rather, although most step-parents are decent, humane, even loving—and some genetic parents are truly despicable—the fact remains that on balance, step-parents are significantly more likely to be "bad parents" than are their genetically connected counterparts.)

In the early stages of courtship, a young, single mother may well be impressed with the care and concern—even, signs of love—that her new boyfriend shows for her child or children. If both man and woman have children from a previous marriage, substantial time and effort are likely to be spent trying to establish harmony among the children as well as between each adult and his or her nonbiological step-child(ren). Sometimes it works, not only during the courtship and honeymoon period, but even later. "Blended" families can succeed, which is just as well, since they are increasingly frequent, because divorce rates have remained high, while growing numbers of men have been ending up with custody of their children.

But sadly and predictably, as any participant in such an arrangement can testify, it is terribly difficult to get blended families to "work." One explanation has long been offered by experts who emphasize the importance of social learning and cultural tradition. They maintain that step-parenting is difficult because the "social role" of step-parent is complicated, ambiguous, and new to the participants. This is reflected in scholarly articles in prominent sociology journals with such titles as "Remarriage as an Incomplete Institution," "Difficulties in the Socialization Process of Stepparenting," and "The Stepparent Role: Expectations, Behavior and Sanctions."[22] The idea behind such "role theory" is that when and if society redefines the role of step-parent, the problem will go away, kind of like Rumplestiltskin, the nasty folktale dwarf who disappeared once his name was discovered and said aloud.

More likely, however, the problem of step-parenting, step-families, blended families, and so forth lies deeper than societal attitudes, and requires more than some benevolent social scientists eavesdropping as Rumplestiltskin accidentally divulges his name.

As evolutionary psychologists Martin Daly and Margo Wilson point out,

> There is a commonsense alternative hypothesis about why some "roles" seem easy and "well-defined" while others are difficult and "ambiguous." It is simply that the former match our inclinations while the latter defy them. Step-parents do not find their roles less satisfying and more conflictual than natural parents because they don't *know* what they are supposed to do. Their problem is that they don't *want* to do what they feel obliged to do, namely to make a substantial investment of "parental" effort without receiving the usual emotional rewards. The "ambiguity"

of the stepparent's situation does not reside in society's failure to define his role, but in genuine conflicts of interest within the stepfamily.[23]

To this, we add that those "genuine conflicts of interest" are genuine because they are, at heart, genetic.

Despite the most sincere efforts by step-parents, the reality is that all too often, the early glow of hoped-for family unity is replaced by simmering embers of resentment, when each parent reverts to norm, which is to say, each wants most for his or her biological offspring. As a result, tensions—from small rumblings to full-sized earthquakes—develop along predictable genetic fault-lines.

Here is a real case study: Ben, a middle-aged physician, wed Alice, a nurse, each of whom had a child from a previous marriage. When they married, Ben's son lived with his mother in a distant state, while Alice's daughter lived with her. Ben liked Alice's daughter, and he felt he, too, could live easily with her and the daughter. In addition, Alice's ex-husband paid generous child support, so Alice's daughter imposed no real financial strain on Ben. Alice also seemed to be fond of Ben's son, and was always nice to him during his visits. At first, all went well.

As the children grew up, it became apparent that Ben was more tolerant of Alice's daughter than Alice was of Ben's son. In fact, when Ben's son wished to come and live with his Dad (and step-mother), Alice was outraged: "How could you possibly expect me to mother your son?" Alice asked. "You work all day. You're gone. Do you know what it will mean for me, to drag him to soccer games and baseball, to cook for a kid who eats like a horse and acts like a pig? This wasn't part of our agreement!"

Step-families often fall apart, for similar reasons. (This one still endures, but just barely.)

Child Abuse, Neglect, and Murder: In Animals

Step-parents have to paddle upstream, struggling against evolutionary pressures to be more genetically selfish than society often expects, trying to be as genuinely altruistic as their spouses desperately wish and as their own disembodied sense of justice might prefer. Sometimes, they capsize: step-parenting is highly associated with child neglect, abuse, and even murder.

Such findings probably comprise the most impressive—and discouraging—evidence for differences between genetic and step-parents. Thus, the sad reality is that step-children are neglected, abused, and killed at rates that are dramatically higher than the corresponding situation for genetic parents and their children. There seems little question that step-parenthood itself is the crucial risk factor.

The evolutionary biology of infanticide becomes clear when we consider how it operates in other animals. After all, human beings have not cornered the market on infanticide. It is widespread in the biological world, even if we exclude those cases of obvious pathology in which, for example, captive hamsters or gerbils occasionally kill and eat their own young. Not surprisingly, when infanticide occurs among free-living animals, the perpetrators nearly always kill infants who are *not* their own.

The path-breaking account of infanticide among free-living animals was by anthropologist Sarah Hrdy, now of the University of California at Davis, who initiated a study of the langur monkeys of India while a graduate student at Harvard. Her findings were initially resisted by many scholars, who refused to accept that animals, in a natural state, behaved so cruelly. But with her work replicated, the basic pattern of infanticide is now undeniable.

It goes as follows. (The details are for langur monkeys, but the general pattern appears to hold for many different species.) Langurs are slender, graceful, silver-gray monkeys at home in trees or on the ground. They live in harems, troops that consist of one male, many females and their offspring. Since there are roughly equal numbers of adult male and adult female langurs, this means that there are many bachelors. These unmated and unruly males gather in loose bands, periodically attempting to invade the troop with its dominant male and numerous females. Every once in a while, they succeed, whereupon after the former dominant male is driven off, fighting breaks out among the victors. One of them eventually emerges as the new boss, drives away his former compatriots, and is left in possession of the reproductive females and their offspring. Now the infanticide begins.

Methodically, and with a most bloodthirsty persistence, the newly ascendant male proceeds to kill the nursing infants as well as any born shortly after his takeover. Their mothers may resist, often with the help of older female relatives (remember kin selection?). But usually in vain. Here is the evil step-parent with a vengeance. In pursuing his bloody business, the infanticidal male langur is not diminishing his own genetic success. Quite the contrary: he is destroying babies fathered by the previous male, in whom he has no evolutionary interest. Moreover, he has a positive reason to get rid of them, because like most mammals, female langurs don't ovulate while they are nursing their young. With their children killed, however, the mothers stop lactating, start ovulating, and mate with the new male—to the advantage of his fitness as well as their own.

In this example, the evolutionary calculus appears especially cold and uncaring, not only when it comes to the infanticidal behavior of the male, but also the wimpy acquiescence of the females, who do not hesitate to copulate with their baby's murderer. But not only is blood thicker than water, it outweighs morality. In this situation, as in so many others, males as well as females are simply doing what maximally contributes to the fitness of their genes.

Hrdy points out, incidentally, that female langurs may not be entirely defenseless. When a male takeover occurs while a female is in the early stage of pregnancy, the female may undergo a "false estrous," in which she actually develops a non-functional sexual swelling and copulates with the newly arrived male. Later, when such females give birth—a month or so "earlier" than usual—the deceived male tolerates these infants, who he presumably treats as his own. Hrdy suggests that the temptation for males to kill infants not their own, combined with female interest in preventing such carnage, may have provided much of the biological impetus for females being sexually receptive to more than one mate. By confusing males as to their fatherhood, females may be able to purchase a degree of infanticidal immunity for their offspring.[24] (This is similar to the suggestion that when female chimpanzees copulate with males outside their group, as described earlier, they do so in order to provide themselves and their offspring with a protection from the lethal "chimpanzee wars" that sometimes take place between neighboring groups.)

Hrdy's path-breaking research cleared away a kind of conceptual log-jam among field biologists. It was followed by a virtual flood of reports on infanticide in other species, including many birds and espcially mammals, from mice to lions.[25] In all such cases, there is a consistent theme, and it is important not to let the dramatic, gruesome trees obscure the evolutionary forest. Following a take-over, infants— offspring of the *ancien régime*—are killed by the newly ascendant male, whose step-parenting is brief, violent, and despicable by human standards, but altogether comprehensible as the outcome of genes jousting with each other for evolutionary success.

There are even practical implications for such fields as wildlife management. Especially when big game is concerned, wildlife officials have long permitted the removal of males while leaving females untouched. Thus, a "buck season," for example, is thought to have relatively little effect on the breeding population of deer or elk (since other males can replace those removed, and a small number of surviving males can inseminate numerous females). But this fails to reckon on the possible role of infanticide by strange males, and thus, the importance of fatherly males in preventing such actions by those newly arrived. A study of brown bears in Scandinavia, for example, has shown that after adult males are removed, mortality is significantly higher among the cubs, who appear more likely to be killed by immigrant males.[26]

Nor is infanticide strictly a "guy thing," although it is indeed more frequent among males than females, probably because males are more often able to enhance their fitness by destroying offspring who are not genetically their own. When the tables are turned, female animals are infanticidal too. Here are three telling examples. The first involves an unusual species known as the wattled jacana. These waterbirds, found in the New World tropics, are noteworthy because they participate in a kind of harem in reverse. Female jacanas are large, bright-colored, pushy and aggressive. They defend territories within which each one mates with several small, drab, retiring males, after which every male tends a separate nest containing the eggs laid by the dominant, territorial female. When a number of female wattled jacanas were experimentally removed, new females moved in, and then—shades of the langur monkeys—they proceeded to kill or drive away the nestlings, which of course were the offspring of the prior female. By so doing, each newly ascendant, territorial female freed up the males in her territory to mate with her, then to rear *her* offspring rather than those of her predecessor.[27]

The second example, also from birds, helps illuminate a kind of avian "fatal attraction," from the popular movie in which a jilted woman seeks to exact lethal retribution against her ex-lover's family. Among several bird species, males are known to achieve a simultaneous sexual liaison with two females, each of whom cares for her own offspring in a different nest. However, the male typically assists only one of these females—the so-called primary mate—to rear her (really, their) off-spring; the secondary mate, forced to be a single parent, is therefore less successful in providing for her nestlings. Investigating this situation among a population of house sparrows in Spain, J. P. Veiga found that secondary females occasionally kill the primary female's off-spring![28]

Among primates, too, adult females sometimes kill another's off-spring. This has been observed almost exclusively among species in which the females disperse; that is, leave the social group in which they were born, and gain entrance into a new group, in which they eventually breed. In such cases, notably gorillas and chimpanzees, an immigrant female may kill the offspring of subordinate, resident females.[29] Such baby-killers, as recent immigrants, are probably unrelated to their victims.

This situation, interestingly, is not the exact inverse of infanticide by male langurs. Unlike the langur case, a newly arrived female gorilla or chimp does not directly increase her mating opportunities by killing infants in her new troop (she also seems unlikely to endear herself to the residents in the process). However, she may well be eliminating potential rivals to her own offspring when she breeds . . . and it is a pretty good bet that eventually, she will.

Child Abuse, Neglect, and Murder: In People

The past is prologue. The beast, too, is prologue. Like it or not, langurs and lions and chimpanzees and gorillas are prologue to an understanding of child abuse, neglect, and murder in human beings.

The Canadian husband and wife team of psychologists Martin Daly and Margo Wilson, professors at McMaster University in Hamilton, Ontario, have pioneered in revealing the evolutionary underpinnings of step-parenting as a risk factor for these unsavory events. Since child-rearing is difficult, costly, and prolonged in our species, natural selection is unlikely to have produced indiscriminate parenting. (If the Mexican free-tailed bat can be fussy about dispensing parental care, so can human beings.) In fact, parental feelings can be expected to vary with the evolutionary interest that children hold for the adults in question: The greater the genetic return, the greater the inclination to invest time, energy and love. The less the evolutionary interest, the less the inclination.

In one study, Daly and Wilson examined whether the household circumstances of children living in Hamilton, Ontario, correlated with victimization rates as measured by child abuse, runaways, and frequency of juvenile offenders. The results:

> Both abuse and police apprehension were least likely for children living with two natural parents. Preschoolers living with one natural and one stepparent were 40 times more likely to become child abuse cases than were like-aged children living with two natural parents. . . . [A]buse risk was significantly higher for children living with a stepparent than for those with a single parent.

The researchers also considered whether other factors besides stepparenting could be responsible for their findings:

> Several variables were examined as possible confounds of household composition. Socioeconomic status, family size, and maternal age at the child's birth were all predictors of abuse risk, but these factors differed little or not at all between natural-parent and stepparent families and could not account for the stepparent-abuse association. As predicted from Darwinian considerations, stepparents themselves evidently constitute a risk factor for child abuse.[30]

Earlier, Daly and Wilson had examined a sample of 177 Canadian households at which the police were called because of an altercation or assault involving a child. When the child had been physically abused, a step-parent lived in the house 48 percent of the time; in cases without abuse, step-parents were present in only 21 percent.[31]

Nor is this pattern unique to Canada. An earlier review of children admitted to New Zealand hospitals for treatment of injuries divided them into two groups, depending on whether their injuries appeared to be accidental or intentionally inflicted. Those in the latter category were more than twice as likely to be living with a step-parent.[32] Research on Pennsylvania families known to be abusive revealed that such abuse is not randomly spewed about, equally likely to strike anyone who happens to be present: Stepfathers typically spared their own genetic offspring, while directing violence toward their step-children.[33]

As impressive as the differences are between step-parents and genetic parents when it comes to child neglect and abuse, they are even more pronounced in the most violent cases; namely, infanticide. In the United States, a child less than two years of age is at 100 times greater risk of being killed by step-parents than by genetic parents![34] In Canada, the risk to step-chldren is "only" 70 times higher.[35] Such findings led the researchers to conclude that "Stepparenthood *per se* remains the single most powerful risk factor for child abuse that has yet been identified."[36]

This phenomenon is so important, and so striking, that it has been subjected to a great deal of scrutiny. As a result, it is now clearly established that the lethal vulnerability of step-children as compared to genetic children is *not* due to reporting bias, to incidental characteristics of people who remarry, to poverty, duration of living together, age of the mother, or number of offspring.[37]

The next step is to compare the actual details of step-parental homicides with those perpetrated by genetic parents. Murder is not only anhorrent but aberrant, no matter who the perpetrator and who the victim. It is especially so when parents kill their offspring. Having shown that step-parents were far more likely than genetic parents to kill their children, Daly and Wilson decided to inquire into those relatively rare cases in which genetic parents did in fact murder their own offspring. The researchers found themselves struck by case reports suggesting that such instances often took place "more in sorrow than in anger," out of what was perceived as unavoidable necessity, and in such misguided desperation that the act of murder was seen by the perpetrators as a rescue rather than a violation.[38]

Here are some of their findings, gruesome and awful, but instructive. Looking at all homicides committed in Canada between 1974 and 1990, Daly and Wilson found that 178 children under five years of age were killed by their fathers and 67 by their stepfathers. Correcting for the proportion of children actually living with genetic fathers versus stepfathers, this means that children were killed by their genetic fathers at a rate of about 6.3 per million children per year; for children living with stepfathers, the rate was about 392 per million per year, *60 times higher*. This is consistent with the whopping difference found whenever such comparisons are made.

Turning next to the way children were killed, beating (hit, kicked, struck with a blunt object) accounted for 82 percent of stepfather murders, compared to 42 percent of murders by genetic fathers. Genetic fathers shot their victims 25 percent of the time whereas stepfathers did so 1.5 percent. "Another way to express these results," write Daly and Wilson,

is to note that whereas a stepfather was about 60 times more likely to kill his preschool child than a genetic father, this contrast does not apply to all means of killing. In particular, a stepfather was not demonstrably more likely than a genetic father to shoot a child, but he was 120 times more likely to beat one to death.

When genetic fathers kill, they tend to do so by shooting or smothering the child, often in its bed; when stepfathers kill, they tend to strike the victim repeatedly. This is consistent with the view that stepfather murders are associated with hostility toward the victim as opposed to a felt "need" to take its life. Even when genetic fathers behave as badly toward their offspring as can be imagined (that is, they kill them) it can be argued that they are still, within the boundaries of their own pathology, acting with a kind of biologically mediated solicitude compared with stepfathers.

This is supported by looking further. When genetic fathers kill their children, they not uncommonly also kill others, including their wives and often themselves as well. Sixty-three of the 178 child-killings were followed by the murderer committing suicide. By contrast, of the 66 stepfathers who killed a step-child, only *one* subsequently killed himself. Genetic fathers therefore seem more likely to kill out of general despondency; stepfathers, out of hostility directed specifically toward their victim. Daly and Wilson also examined data collected from England and Wales. In these countries, the population is about twice as large and the homicide rate approximately one-half that of Canada. But the same pattern persisted. (The data on homicidal step-mothers vs. genetic mothers are too scanty at present for conclusions, but no one should be surprised if a pattern comparable to that for step-fathers vs. genetic fathers is eventually revealed.)

Daly and Wilson conclude their research article by noting that

> stepfathers do not merely kill children at higher rates than genetic fathers. They kill them in different ways, and for different reasons. The different attributes of these two categories of homicides support the hypothesis that the risks to children in stepfamilies reflect predictable differences between stepparental and genetic parental solicitude.

The similarity with animal infanticide suggests that child abuse, neglect, and murder might be evolved adaptations, if step-parents are following a strategy designed to eliminate nongenetic children so as to increase the success of their own genetic offspring. (Abuse and neglect would thus represent incomplete or unsuccessful efforts toward this generally unconscious goal.) Given that it is adaptive for gene-sharing to correlate with love, care-taking, and a general willingness to invest in a child's welfare, then a lack of genetic relationship may correlate, in turn, with a comparable lack of parental love, care-taking, and watchfulness, leading in extreme cases to a degree of disinhibited aggression and violence. After all, despite its many rewards, child-rearing can certainly be stressful, even for the most well balanced and devoted parents. It is understandable—if not pardonable—that without genetic connection to ameliorate the rough edges, there would be a lower threshold for adults' ability to tolerate infant crying, children's interrupting, and the normal demands of even the most well-behaved youngsters. For unstable adults already teetering on the edge of self-control, step-parenthood could well make a tragic difference.

As we have seen, kin selected altruism operates strongly even when it is not reciprocated, when the altruists keep on giving and the recipients keep on benefiting. In such cases, both parties end up better off (more fit, in the strictly evolutionary sense), and so, both can be expected to be satisfied with such an arrangement. But in the absence of shared genes, altruism nearly always carries with it the expectation of reciprocity. Parenting is an extreme form of kin selection, with the one-directional flow of benefits exaggerated by the fact that parents are so much older, larger, wiser, and more powerful than their children. Parents are biologically (and therefore psychologically as well as socially) expected to conform to these expectations and to accept a largely one-sided relationship with their children.

Step-parents, on the other hand, often struggle to mimic genetic parents, in their behavior as well as their feelings. And yet, even for those who are well-adjusted, loving, and nonviolent, it is a difficult undertaking. The most well-intended advice, anecdotes, and pop psychology—even when utterly non-Darwinian—acknowledges that step-parenting is difficult, as is being a step-child. Evolutionary thinking, however, has not made impressive headway into the traditional wisdom of social sciences applied to step-parenting, which, as we have said, still tends to attribute its near-universal stress to problems of "role definition," and "social expectations." It is far more likely that the fault lies not in our stars and not in society but in ourselves, that is, in the fact that we are biological creatures, carrying on a long-standing tradition by which genes struggle with other genes. And all too often—unless we watch our step—our bodies simply go along.

Notes

1. U. Seibt and W. Wickler. 1987. Gerontophagy versus cannibalism in the social spiders *Stegodyphus mimosarum* Pavesi and *Stegodyphus dumicola* Pocock. Animal Behaviour 35: 1903-1905.
2. Do spiders have flesh?
3. P. Gagneaux, David Woodruff, and Christophe Boesch, 1997. Furtive mating in female chimpanzes. Nature 387: 358-359.
4. Gjershaug, J. O., T. Jarvi, and E. Roskaft. Marriage entrapment by 'solitary' mothers: A study on male deception by female pied flycatchers. American Naturalist 133 (1989): 273-276.
5. C. Day Lewis. 1992. The Complete Poems. Stanford University Press: Stanford, CA.
6. G. F. McCracken. 1984. Communal nursing in Mexican free-tailed bat colonies. Science 223: 1090-1091.
7. T. D. Williams. 1994. Adoption in a precocial species, the lesser snow goose: Intergenerational conflict, altruism or a mutually beneficial strategy? Animal Behaviour 47: 101-108.
8. M. D. Beecher. 1982. Signature systems and kin recognition. American Zoologist 22: 477-490.
9. M. Daly and M. Wilson. 1982. Whom are newborn babies said to resemble? Ethology and Sociobiology 3: 69-78.
10. J. M. Regalski and S. J. Gaulin. 1993. Whom are Mexican infants said to resemble? Monitoring and fostering paternal confidence in the Yucatan. Ethology and Sociobiology 14: 97-113.
11. N. J. S. Christenfeld and E. A. Hill. 1995. Whose baby are you? Nature 378: 669.
12. W. T. DeKay. 1995. Grandparental investment and the uncertainty of kinship. Paper presented to the 7th annual meeting of the Human Behavior and Evolution Society, Santa Barbara, CA.
13. H. A. Euler and B. Weitzel. 1996. Discriminative grandparental solicitude as reproductive strategy. Human Nature 7: 39-59.
14. A. Pashos. 2000. Does paternal uncertainty explain discriminative grandparental solicitude? A cross-cultural sutdy in Greece and Germany. Evolution and Human Behavior 21: 97-109.
15. R. C. Sargent. 1989. Allopaternal care in the fathead minnow, Pimephales promelas: Step-fathers discriminate against their adopted eggs. Behavioral Ecology and Sociobiology 25: 379-386.
16. Harry S. Power. 1975. Mountain bluebirds: Experimental evidence against altruism. Science 189: 142-143.
17. N. B. Davies, B. J. Hatchwell, T. Robson, and T. Burke. 1992. Paternity and parental effort in dunnocks *Prunella modularis*: how good are male chick-feeding rules? Animal Behaviour 43: 729-745.
18. T. Hasegawa and M. Hiraiwa. 1980. Social interactions of orphans observed in a free-ranging troop of Japanese monkeys. Folia Primatologica 33: 129-1598.
19. H. Kummer. 1968. Social Organization of Hamadryas Baboons: A field study. University of Chicago Press: Chicago.
20. Mark V. Flinn. 1988. Step—and genetic parent/offspring relationships in a Caribbean village. Ethology and Sociobiology 9: 335-369.
21. Stith Thompson. 1955. Motif-Index of Folk Literature. Indiana University Press: Bloomington.

22. A. Cherlin. 1978. Remarriage as an incomplete institution. American Journal of Sociology 84: 634-650; D. R. Kompara. 1980. Difficulties in the socialization process of stepparenting. Family Relations. 29: 69-73; and J. Giles-Sims. 1984. The stepparent role: expectations, behavior and sanctions. Journal of Family Issues 5: 116-130.

23. Martin Daly and Margo Wilson. 1988. Homicide. Aldine de Gruyter: Hawthorne, NY.

24. S. Hrdy. 1979. The Langurs of Abu. Harvard University Press: Cambridge, MA.

25. F. S. vom Saal, and L. S. Howard. 1982. The regulation of infanticide and parental behavior: implication for reproductive success in male mice. Science 251: 1270-1272; A. E. Pusey and C. Packer. 1992. Infanticide in lions. In Infanticide and Parental Care. S. Parmigiani, F. vom Saal, and B. Svare, eds. Harwood Academic Press: London.

26. J. E. Swenson, F. Sandegren, A. Sodergerg, A. Bjarvall, R. Franzen, and P. Wabakken. 1997. Infanticide caused by hunting of male bears. Nature 386: 450-451.

27. S. T. Emlen, N. J. Demong, and D. J. Emlen. 1987. Experimental induction of infanticide in female wattled jacanas. Auk 106: 1-7.

28. J. P. Veiga. 1990. Infanticide by male and female house sparrows. Animal Behaviour 39: 496-502.

29. Goodall, J. 1977. Infant killing and cannibalism in free-living chimpanzees. Folia Primatologica 28: 259-282 ; D. Fossey. 1984. Infanticide in mountain gorillas (Gorilla gorilla beringei) with comparative notes on chimpanzees. In G. Hausfater and S. Hrdy, eds. Infanticide: Comparative and evolutionary perspectives. Aldine de Gruyter: Hawthorne, NY.

30. Martin Daly and Margo Wilson. 1985. Child abuse and other risks of not living with both parents. Ethology and Sociobiology 6: 197-210.

31. M. Daly and M. Wilson. 1981. Child maltreatment from a sociobiological perspective. New Directions for Child Development 11: 93-112; see also M. Daly and M. Wilson. 1996. Violence against stepchildren. Current Directions in Psychological Science 5: 77-81.

32. D. M. Ferguson, J. Fleming, and D. P. O'Neill. 1972. Child Abuse in New Zealand. New Zealand Government Printing Office: Wellington.

33. J. L. Kightcap, J. A. Kurland, and R. L. Burgess. 1982. Child abuse: A test of some predictons from evolutionary theory. Ethology and Sociobiology 3: 61-67.

34. M. Daly and M. Wilson. Homicide.

35. Martin Daly and Margo I. Wilson, 1988. Evolutionary social psychology and family homicide. Science 242: 519-524.

36. M. Daly and M. Wilson, Homicide.

37. E. Voland. 1988. Differential infant and child mortality in evolutionary perspective: Data from late 17th to 19th century Ostfriesland. In Human Reproductive Behavior. L. Betzig, M. Borgerhoff-Mulder and P. Turke, eds. Cambridge University Press: Cambridge; M. Wilson and M. Daly. 1987. Risk of maltreatment of children living with stepparents. In Child Abuse and Neglect: Biosocial dimensions. R. J. Gelles and J. B. Lancaster, eds. Aldine de Gruyter: Hawthorne, NY.

38. Martin Daly and Margo I. Wilson. 1994. Some differential attributes of lethal assaults on small children by stepfathers versus genetic fathers. Ethology and Sociobiology 15: 207-217.

6

Conflict between Parents and Offspring

"The course of true love never did run smooth." Directed to romantic love between adults, this observation from Shakespeare's *A Midsummer Night's Dream* also applies to parents and offspring.

It is more than a bit surprising. Thus, there is some logic, for example, to conflict among men over status, dominance, or money, all of which translate ultimately into reproductive opportunities with women, and in turn, social and biological success. Similarly for conflict among women, although usually with less bluster and violence. There is even some evolutionary rationale to battles between the sexes, insofar as men and women are different—genetically as well as in their reproductive tactics and strategies—despite the fact that their interests converge in reproduction. But the evolutionary interests of parents and their offspring would seem to coincide perfectly, since parents want their children to succeed, as do the children themselves. And so, little or no conflict might be expected.

Years of Walt Disney "True Life Adventures," combined with cartoon images from *Bambi* and *Dumbo* to *Lady and the Tramp* and *One Hundred and One Dalmations* have both reflected and generated the expectation that animal parent and child—especially, mother and child—are the epitome of shared goals and perfect amiability. The image among human beings is, if anything, even more clearly established: Madonna and Child convey a sense of peace and contentment that transcends the merely theological.

When rough spots emerge in the parent-child nexus, the traditional view among psychologists and sociologists has long been that the culprit is simply misunderstanding, with its attendant failures of communication. Everyone means well. It is just that in the course of conveying heartfelt (shall we say, altruistic?) parental assistance, advice, and information to the child, sometimes there are problems, largely because the child—being young—is inexperienced, perhaps occasionally a bit headstrong, and generally uninformed as to where its true interests lie. The smart child, the well-socialized child, the gradually more mature child, eventually recognizes that its best interests lie in going along with parental inclinations, at which point conflict ceases and "socialization" has been achieved. According to this view, when conflict arises between parent and offspring, it is largely due to the fact that children are primitive, even barbaric little creatures, who need time to become incorporated into the society of responsible adults.

Then there is the psychoanalytic tradition, which focuses on sexual rivalry, especially between sons and fathers. As a result of presumed Oedipal conflict, boys are terrified that their fathers will castrate them, girls resent not having a penis, and so all hell breaks loose until eventually things quiet down because the child gets older and less obstreperous, reconciling its primitive conflicts by assuming the social role appropriate to its station.

The view from revolutionary biology is quite different, rather darker, and much more persuasive.

Theory of Parent-Offspring Conflict

We are once again indebted to Robert Trivers, who also provided the first insightful discussion of reciprocal altruism. When Thomas Huxley first read Darwin's *Origin of Species*, he supposedly exclaimed, "How stupid of me not to have thought of that!" Trivers has been in the habit of evoking similar responses from modern biologists, over things that are equally "obvious"—after he points them out! In a landmark analysis published in 1974, Trivers pointed out that the conventional wisdom regarding parent-offspring relations is all wet.[1]

Trivers emphasized that although parents and offspring do indeed have a substantial shared interest, the overlap is not complete. As to the shared interest, it is based on a 50 percent probability that any gene present in a parent is also present in the child. This, as we have seen, is the biological basis of reproduction itself, and it leads to a powerful common goal: Insofar as the child succeeds, the parent does, too. Or at least, one-half of the parent. But just as there are often two sides to every story, there is also the other half when it comes to each offspring's genetic make-up. And this half is not shared between any given parent and child. When it comes to parent-child relations, biologists have in a sense been focused on the genetic glass half full (the amount shared between parent and offspring), all the while ignoring the other, empty half, that part of child-parent genetic identity that does *not* exist.

Not only that, but biologists—and most psychologists and sociologists to an even greater extent—have treated the child as an appendage to the parent, rather than a separate being with its own strengths as well as weaknesses, and, even more important, its own agenda.

Consider a newborn infant, say, an elephant calf (or, for that matter, a human being). Initially, the interests of the infant and its mother coincide: The infant needs various things from its parent, milk in particular. And the mother is prepared to meet these needs. In the short term, her hormones as well as her anatomy predispose her to lactate; in the long run, her evolutionary interests are also served by helping to make a healthy, successful child. Everything is just fine. Mother and child agree.

But then, gradually, something happens. The infant grows older, larger, and less needy of its mother's milk. At the same time, the mother becomes inclined to discontinue nursing. After all, milk is energetically costly to produce, and at some point, the mother will do better in terms of her own fitness if she stops investing in her current child and prepares to put precious resources into another. (In most mammals, as already stated, lactating females are inhibited from ovulating, so the nourishing of one offspring literally precludes making another.) This, in itself, need not lead to conflict. If the infant agrees with the mother, then weaning would be as smooth as the early stages of nursing. Unfortunately, it usually isn't.

The mother, we must recall, is ultimately interested in making the most of *her* fitness, not necessarily that of her offspring. In fact, her only reason for creating that offspring in the first place was as a means of advancing her own fitness. By the same argument, the infant is ultimately interested in making the most of *its* fitness, not necessarily that of its mother. Another way of looking at it: The infant is 100 percent related to itself but only 50 percent related to its mother. (And vice versa for the mother.) As a result, the infant devalues its mother's interests by a factor of one-half. When the mother decides that the balance of costs and benefits—for *herself*—favors an end to nursing, the infant can be expected to see things differently. After all, baby elephant (or human) is only one-half as concerned for its mother's cost/benefit considerations as the mother is.

Think of the therapist's cliché, "I can really feel your pain." The infant can feel only one-half the mother's pain. And vice versa for the mother.

The upshot is that after a period in which mother and infant are in agreement, with both enthusiastic about nursing, there arises a zone of conflict, in which the mother wants to discontinue while the infant feels otherwise. Time passes. During this difficult, conflictual period, the mother and infant are locked in a battle of evolutionary wills, with the infant selected to demand more than the mother is selected to give. But there is light at the end of the tunnel. For the mother, the cost of nursing continues to mount, while for the infant, the benefit of nursing begins to decline. At some point, therefore, their interests once again coincide. Even with the infant devaluing its mother's costs by a factor of 1/2, eventually it is also in the infant's interest if the mother stops giving so much, and starts taking care of herself. (One way of looking at it: the infant wants the mother to provide it with siblings.) The mother is only too happy to oblige and so the two parties agree at last, and nursing is finally discontinued.

Here, then, is the basic pattern: early agreement, followed by conflict in which the infant wants more than the mother wants to give, and then eventual reconciliation. It is widely found among living things.

Examples of Parent-Offspring Conflict

Take a cat with kittens. When her offspring are young, the mother initiates most of the nursing bouts, until the young are around twenty days of age. Then, things begin to shift. Between twenty and thirty days, kittens and mother are equally likely to initiate nursing. By about day thirty, however, the momentum has changed so much that it is the kittens who attempt to nurse, while their mother discourages their efforts, to the point that she is likely to get up and leave, something that happened only rarely when the kittens were newborn. In one series of observations, a mother cat regularly escaped from her increasingly pesky offspring by jumping onto a shelf they could not reach.[2]

The evolutionary genetics of parent-offspring conflict provides a new way to look at this widespread phenomenon known as weaning conflict. It is found in virtually all mammals, human beings not the least. Something similar even takes place among birds. Large nestlings—big enough to fly, hence known as fledglings—can often be found pursuing their harried parents, importuning them for food. In late spring throughout North America, it is common to see these fledglings quivering their wings and uttering incessant "begging" calls, while the parents back away, look far into the distance (as though trying to ignore what is in front of them), and often literally take wing, pursued by their nearly grown, but indefatigably demanding offspring.

One of the most comical examples is the so-called "feeding chase" of flightless Adelie penguins, in which the adults waddle about all over the rookery, desperately pursued by rapacious juveniles who are hoping to get their worn-out parents to regurgitate just one more chunk of precious, predigested fish.

Conflict over weaning (or its equivalent in birds) does not exhaust the potential for parents and offspring to disagree. In this way of looking at parent-offspring relations, the root of all evils is conflict over "parental investment," anything that parents provide to their offspring and that contributes to the offspring's success, but which carries with it a cost as measured by the parent's ability to produce and invest similarly in additional offspring.[3] Parental investment is anything beneficial to the offspring but costly for the parents to provide. Aside from milk, it includes other types of food, defense against predators, time spent giving instruction, keeping the children warm, tending to the nest or den, and so forth.

When a rhesus monkey is just born, the infant stays close to its mother. When the baby wanders off, nearly always it is retrieved by the mother. By about fifteen weeks or so, however, the tables have turned, and it is the young monkey who seeks to initiate contact, only to be increasingly rebuffed by the mother.[4]

The transition from conflict to offspring independence, although rarely smooth, is often accomplished gradually. Maybe this is because parents are selected to minimize the stress felt by their offspring, or because with parent and offspring so intensely engaged in the struggle, neither party can win abruptly and cleanly. A European species of woodland bird, the spotted flycatcher, reveals a nice example of conflict combined with a gradual transition to offspring independence.

During the first nine days after hatching, spotted flycatcher parents feed their young regardless of whether these offspring are silent or calling. After ten days, feeding only takes place when the young call vigorously. Chicks also begin chasing their parents, demanding their food, which parents are increasingly reluctant to provide. Increasingly, as time goes on, food is *only* given up after a chase, and sometimes not even then. By day eighteen, only 20 percent of feeding chases end in the offspring cadging a meal. The actual size of the meal goes down as time goes on, despite the fact that larger nestlings presumably need more food, not less.

By this time, not surprisingly, the offspring are beginning to get their own nourishment. The outcome—whether directly intended or not—is that growing spotted flycatchers have been gradually induced to earn their own living: Initially, they were fed, whatever they did. Then, they had to call. Then, they had to chase their parents, ever more determinedly, and for a longer time. In the end, they had no choice but to forage for themselves. The offspring became better at catching their own food while still being provisioned by their parents. So they had a "safety net" of sorts. But eventually, like it or not (and to look at them, mostly the answer is "not"), the young became independent, and the parents, off the hook.[5]

It is not hard to draw parallels to people. In fact, it is hard not to do so! Among human beings, parental investment includes money but is not limited to it. It also goes on longer, and is more intense, than in any other species. Many a harried parent, struggling to provide for even the most rewarding child, will answer the question, "What do you want your child to be?" with an immediate reply: "Self-supporting!"

Teenagers are often encouraged to work part time, in the evenings or weekends, if they want extra spending money, or a car. Then there is college. Some parents pay for it willingly; others, only grudgingly. Yet others, not at all. Clearly, as the offspring grow older, some sort of transition is reached, and most of the time, not surprisingly, children would like more parental investment, continued longer, than parents are inclined to provide. It would be interesting to see whether this pattern holds cross-culturally.

Conflict Concerning Third Parties

Parent-offspring conflict has some surprising implications. Robert Trivers pointed out, for example, that parents and offspring can be expected to disagree over many things, beyond the relatively simple question of how much parents should invest in a child, or whether they should do so at all. They can be expected to disagree, for instance, over the child's behavior toward a third party. In the most obvious case, and probably the most frequent, this third party is a brother or sister. For a parent, who is equally related to each child, every one is equally important. (To this, add the crucial caveat "all other things being equal." It is assumed that each child is equally likely to be successful, and also that every one is equally needy; in other words, they are all comparable in their ability to convert parental investment into fitness. This is never true in the real world, but is essential to any basic model.)

Parents benefit any time one of their children acts altruistically toward a brother or sister. So long as the benefit to the recipient is greater than the cost to the altruist, the parent comes out ahead. For example, imagine little Jimmy wants some food that little Suzie is eating. If Jimmy's need for the food is greater than the cost to Suzie in giving it up, then not only is Jimmy benefited by the gift, but so is the parent. But what about Suzie?

Well-fed Suzie, too, can benefit by giving food to hungry Jimmy, as we explored when we looked at kin selection. But—and it is a crucial "but"—her calculation is different from her mother's. Whereas Mom wants Suzie to help Jimmy any time Jimmy's gain exceeds Suzie's loss, Suzie drives a harder bargain: she wants to help Jimmy only when Jimmy's gain is *twice* as large as Suzie's loss. This is because Suzie is genetically related to Jimmy by a factor of 50 percent, but is 100 percent related to herself! In other words, she is expected to value herself twice as much as she values her brother.[6] Once again, this is a replay of altruism via kin selection and the maximization of inclusive fitness. Its relevance to parent-offspring conflict arises when another party—the parent—is introduced, and then we can expect parent and offspring to disagree over whether altruism is called for, and how much.

With equal justice, the issue can be turned on its head: Siblings will be inclined to act selfishly toward each other whenever the cost to the recipient (i.e., the victim) is less than twice the benefit derived by the initiator. By contrast, the parents would be opposed to any such behavior, so long as the cost incurred by the victim is greater than the benefit derived by the selfish sibling. In between—that is, whenever the cost of sibling selfishness is greater than its benefit, but less than twice that benefit—parents and offspring can be expected to disagree, with parents wanting offspring to refrain from selfishness and offspring wanting to indulge themselves.

There is a common phrase for this whole business: sibling rivalry.

It is also a common parent-offspring experience: parents urging a child to play more nicely with its sibling than the child itself wants to do. Or urging a child to share when the child is inclined to be selfish. Part of the power of the evolutionary approach comes from its recognition that such cases are not simply due to stubbornness or sheer perversity on the part of a rivalrous or selfish child. Rather, the key is that children are inclined to act in response to *their own* biological interests, rather than those of their siblings, their parents, society, or anyone, for that matter.

In most such cases, the bottom line is that parents are expected to exhort, extort, or otherwise try to induce their offspring to act more altruistically, more benevolently, more pro-socially than the offspring would choose, if left to their own devices.

Studies taking a cross-cultural perspective on parent-offspring conflict have barely begun. The possibilities, however, are immense. To give but one example: Recall the phenomenon of "helpers at the nest" among birds, in which young adults sometimes remain in attendance at their parents' nest, garnering an inclusive fitness benefit via enhanced breeding success on the part of their parents. A possible human analogy may well exist, in that it is fairly common in certain social groups for one child within each family to remain unmarried and serve as "nanny" to enhance the eventual reproductive success of his or her siblings. It would be worthwhile to examine these cases for possible signs of parent-offspring conflict, since even though such nannies may reap a kin selected payoff, they would presumably do even better by rearing their own families.[7]

For another example, consider behavior toward cousins. Parents are more closely related, genetically, to their nieces and nephews than cousins—offspring of two siblings—are to each other. (Uncles and aunts have a genetic relationship to their nieces and nephews of 1/4, whereas cousins are only one-half as related, with a coefficient of 1/8.) As a result, it can be predicted that parents who are full siblings would like their respective children to be twice as nice to each other as the cousins are inclined to be!

It is very common, at least in the United States, for cousins to spend time together. But this is notably true when the cousins are young, at which time their activities are largely controlled by their parents. As time goes on, the extended family grows older, and parents become less influential and eventually die. At this point, no longer subject to parental pressure, cousins typically drift apart, or at least, they tend to become less close than in the past, when they were subject to their parents and less able to assert their own interests.

When it comes to parent-offspring conflict over behavior involving a third party, step-parenting is likely to take the cake. If parents and their children are expected to disagree over the latter's behavior toward siblings, or cousins, at least in these situations the parents and offspring have some shared genetic interest. But when a step-parent is involved, conflict can be sky-high. The genetic parent is in a bind: he or she would like the child to behave at least somewhat benevolently toward the step-parent and the step-parent's children (if any), especially if the new husband and wife are young enough to contemplate having their own children together. At the same time, parents are likely to realize—at some level—that their genetic offspring's interests are quite different.

The child, once again, sees things in its own way. Even if the step-parent does not represent a lethal or even dangerous threat (and in the overwhelming majority of cases, no such threat exists), and even though the child often has a genuine interest in the happiness of the genetic parent, the truth is that the genetic child has relatively little biological interest in the welfare of its step-parent, or the step-parent's children from a previous union. Under these conditions, we can predict a virtual free-for-all of parent-offspring conflict, in which issues, previously quiescent—or suppressed—during courtship, are likely to resurface as the outcome is renegotiated.

There is also more than a little opportunity for conflict between the spouses, with each inclined to favor dispensing resources preferentially toward its genetic offspring rather than toward the step-children.

Siblicide

Then there is the curious case of "siblicide," the killing of one sibling by another. It didn't start with Cain: Animal examples are well known, sometimes beginning remarkably early in life. It is not uncommon, for example, that a newborn hyena will kill its brother or sister. Baby hyenas—of both sexes—are born with fearsome and fully developed teeth, as well as a truly murderous disposition. The youngsters vie to literally tear each other apart, presumably so as to monopolize food from the mother.

It seems likely that parent hyenas would rather their offspring be mutually respectful, or at least, something less than mutually murderous. If so, then hyena siblicide qualifies as an example of parent-offspring conflict. There are other interesting cases, however, in which parents seem at best indifferent to whether their offspring kill each other. In fact, they sometimes appear to be indirect accomplices in the grisly business.

The best-studied examples come from the work of Douglas Mock, zoologist at the University of Oklahoma. By studying great egrets and cattle egrets—long-legged wading birds of ponds and marshes—Mock has documented that among these animals, older siblings commonly kill the younger, while the parents stand idly by. Even as nestlings, these birds are the avian equivalents of hyena cubs, outfitted with long, sharp, saber-like beaks, with which they typically engage in prolonged and lethal combat, four or five battles per day, some of them consisting of more than 100 vicious thrusts, parries, and counter-thrusts.

Just as a simplistic view of natural selection would not suggest that parents and offspring should be in conflict, parental indifference to siblicide seems to run counter to evolutionary expectation. Indeed, as we have seen, although it may occasionally be in the interest of an offspring to be nasty—even murderous—toward its sibling, such behavior should run counter to the interest of the parent, and be prevented when possible. And yet, in nearly two decades of observing almost 3,000 battles among young egrets, Mock never saw a parent intervene! Why not?

Egrets lay three eggs. The first two that hatch enjoy an advantage, if only because they are larger than the third; typically they collaborate to kill their younger sibling. From the parents' perspective, it appears that the third egg is an insurance policy, likely to survive if its older sibs happen to die. In addition, if food supplies are exceptionally abundant, parent egrets may be able to rear all three offspring. Under normal circumstances, however, it seems that parent egrets not only permit their older offspring to kill the younger, they actually encourage it. Thus, Mock has found that parents collaborate in the siblicide by endowing their first two nestlings with an enormous dose of testosterone, up to twice the amount provided to the third nestling. The juiced-up elder sibs then employ their enhanced aggressiveness to the detriment of the youngest, but to their own—and their parents'—genetic benefit.[8]

Egrets are an extreme example of parents loading the dice in favor of certain offspring. Among some species, parents do the opposite: Instead of playing favorites, canaries, for example, provide the greatest share of testosterone to the youngest hatchlings, thereby leveling an otherwise lethal playing field. Biologists' understanding of siblicide is, to coin a phrase, in its infancy. For every family of murderous little brutes, there may be hundreds or thousands composed of peaceable brothers and sisters. It also remains to be seen whether the canary or egret pattern is the more common, and what ecological factors have resulted in one strategy as opposed to its alternative.

It is undeniable, however, that siblicide is more frequent than previously thought. Embryonic sharks, for example, begin their predatory lives with an early burst of sibling-sibling competition, devouring each other in utero as they swim about before they are born. And pronghorn antelope fetuses kill each other in the womb, presumably thereby increasing the amount of mother's milk that they will eventually obtain. Siblicide has even been reported for plants. The Dalbergia tree of India, to take one case, disperses its seeds via pods that float in the wind; lighter pods travel farther. The first Dalbergia seed to develop produces chemicals that kill its pod-mate siblings, thereby giving it sole occupancy of the vehicle, and thus, the advantage of greater dispersal distance.

Fortunately, there is no evidence—as yet—that human beings partake of siblicide as an evolutionary strategy. But it is not impossible. Even though *Homo sapiens* normally gives birth to one child at a time—and thus might seem spared the temptations of young egrets to kill their nest-mates—human offspring remain dependent for many years, providing opportunity for selection to have created various subtle, but potentially lethal strategies. There is, for example, something known as the "vanishing twin syndrome," whereby many pregnancies that begin with twins end up with a singleton child being born. Until recently, such early embryonic weeding-out has been attributed to "spontaneous abortion," but in a sense, nothing in nature is truly spontaneous. The word is simply a cover-up for our ignorance of the true causes. The possibility at least exists that human embryos—with or without the active if unconscious collusion of their mother—exert a nefarious and violent evolutionary strategy upon each other.[9]

Finally, the mere threat of one sibling killing another could serve as a kind of evolutionary blackmail, inducing parents to provide resources in excess of what might otherwise be forthcoming. Imagine that you are a nestling bird, compelled to share with your nest-mates any insects brought by your parents. Granted, your parents are probably hard-working, but given the likely reality of parent-offspring conflict, it is not unreasonable that perhaps they are holding out on you (and your sibs), if only just a little bit. Maybe they occasionally eat a mayfly or two themselves, when they could have brought it to you; or maybe they are just prone to slack off now and then. In any event, insofar as you could profit from getting more food than you are currently receiving, you might have a plausible evolutionary rationale for doing away with one or more of your sibs (both for the nutrition they provide and also to get a larger share of the available goodies). This credible threat—that you might resort to siblicide—could drive your already hard-working parents to work even harder, in a sense hoping to buy you off.[10]

Strategies and Counter-Strategies

Given that parents and offspring are likely to be in conflict, at least sometimes, it seems pretty obvious who would win, at least most of the time: the parents. They are bigger, stronger, older, wiser, and more experienced in the ways of the world. Offspring shouldn't have a chance.

As Trivers pointed out, infants can hardly fling their mothers to the ground and nurse at will. And yet, they are not entirely helpless, because, ironically, their vulnerability outfits them with an arsenal of weapons. Due to their physical inferiority relative to their parents, we can expect, in fact, that the weapons employed by offspring would be almost entirely behavioral. They have become experts at psychological warfare.

Don't forget that parents and their offspring have a significant shared interest as well as a predicted pattern of conflict. In the long run, it is almost as much in the parent's interest to have its offspring succeed as it is in the offspring's interest to do so. (That is why the time of parent-offspring conflict is sandwiched between two periods of parent-offspring agreement.) Parents—especially among birds and mammals—generally have something of value to convey to their offspring: food, protection, lessons of various sorts, and so on. And since offspring know, better than the parents, what their real needs are, and how severe they might be, it is in the parents' interest to be attuned to these needs, so as to respond appropriately. Unless in the grip of especially intense parent-offspring conflict, parents respond when offspring cry, or when they give other indications of distress. Put it this way: any genetic tendency on the part of parents to be indifferent to their offspring's requirements would likely be selected against. This, in turn, gives the offspring just the opening they need. They can proceed to manipulate—or attempt to manipulate—parental gullibility by sending false signals: pretending to be more needy than they really are, so as to get that extra amount of parental investment they desire, and that parents are reluctant to provide.

At the same time, the opposite possibility also exists: that infants have been selected to send honest signals to their parents. When food is scarce or if they are in poor condition themselves, parents could be tempted to cut their loses and cease investing in costly children, especially if they might have an opportunity to breed again some other day. In this case, children would be under evolutionary pressure to send potentially reassuring signals to their parents, containing information that says, in effect, "I am healthy and strong. Therefore, I won't take much from you, and moreover, I will probably provide a good evolutionary return on your investment." Given the choice, parents would likely favor offspring whose signals of this sort are especially likely to be honest, that could not readily be faked. In short, selection should favor offspring signals that are expensive to produce, since offspring in poor condition would necessarily be less able to send signals of this sort.

In this respect, it is interesting that crying by human infants is quite costly, involving an expenditure of calories that is about 11 percent higher than a quiet baby. In addition, various illnesses and debilitating conditions produce consistent variations in "cry characteristics," variations to which parents respond. Crying, in short—especially on the part of healthy babies—may be a way of proclaiming that one is worthy of parental investment.[11]

Children don't only cry and complain. They also smile when they are happy, that is to say, when their needs are met. Not surprisingly, adults find this gratifying, to the point of performing all sorts of peculiar antics in the hope of generating a smile from a young child. Children could thus withhold their smile until their needs have been satisfied. Similarly with sleep. For exhausted, sleep-deprived parents, sometimes the greatest reward their infant can provide is to go to sleep. Almost certainly, the child is not saying to itself, "Don't go to sleep until all of your needs are met," any more than it is plotting "Hold off on that smile until you get a bottle, or a breast, or a goofy look from Uncle Charley." But it would make biological sense if that same child were to feel agitated and restless until its (unconsciously) desired outcome is achieved, and to act accordingly.

One extreme of agitation and restlessness is the temper tantrum, in which a child behaves in a way that is not only unpleasant for the parent, but in which the child even threatens to injure itself. Picture a child beating its head against the wall unless or until it gets its way, in a game of chicken played against its own parents.

Temper tantrums, or their equivalent, have even been described for animals. Here is wildlife biologist George Schaller's description of white pelicans in Yellowstone National Park:

> Young, ten or more days old, often begged vigorously for their food. Usually a young pelican sat very upright in front of its parent, with neck stretched high and wings beating, until it was admitted to the pouch. [allowed to feed] Sometimes, however, a young bird ran to an adult, threw itself on the ground, and beat its wings wildly, all the while swinging its head from side to side. Occasionally the young lay on its side, beat one wing, suddenly jumped up, ran at and pecked several young in the vicinity, driving them away, only to continue begging. It also grabbed, shook, and bit its own wing with the bill as it turned its body around and around, growling all the time. In the words of Chapman (1908, pg. 102) the young "acts like a bird demented."[12]

And here is Jane Goodall, recounting the trials and tribulations of parent-offspring interactions among chimpanzees:

> Temper tantrums are a characteristic performance of the infant and young juvenile chimpanzee. The animal screaming loudly either leaps into the air with its arms

above its head or hurls itself to the ground, writhing about and often hitting itself against surrounding objects. The first temper tantrum observed in one infant occurred when he was 11 months old. He looked around and was unable to see his mother. With a loud scream he flung himself to the ground and beat at it with his hands, and his mother at once rushed to gather him up. Two infants showed tantrums in connection with weaning and this has been recorded also in infant baboons and langurs. . . . Yerkes (1943), when describing tantrums, comments that he often saw a youngster "in the midst of a tantrum glance furtively at its mother or the caretaker as if to discover whether its action was attracting attention." In captivity, individuals are less prone to indulge in temper tantrums as they grow older, and this was also true of wild chimpanzees.[13]

It seems likely that parent pelicans and chimpanzees are as irritated and perplexed by their offspring's tantrums as human parents are. Probably, they are also about as likely to give in.

For their part, parents would be selected to distinguish indications of real need from those that are exaggerated and dishonest. As the folk-tale advises, it may be dangerous to "cry wolf," that is, to insist on being needy when in fact, you are merely greedy. Parents can learn to see through such exaggerations, but they are unlikely to discount them altogether. When in doubt, it may pay them to err on the side of leniency and generosity, especially if the cost to the parent is low and the outcome might be real offspring distress or even death if the child is truly in trouble.

It can also be predicted that older parents would be better at discriminating whether offspring solicitations spring from genuine need or from parent-offspring conflict, if only because they are more experienced.

Trivers proposed that the psychological phenomenon of "regression" may have its roots in the parent-offspring tug of war, since younger offspring are usually more in need, less likely to be in conflict with the parent's self-interest, and thus, most likely to send signals to which parents respond positively. Most people find babies and toddlers attractive. Also puppies, kittens, and so on: "The only thing wrong with a kitten is that/ It grows up to be a cat." So wrote Ogden Nash, giving words to a nearly universal human weakness for animals that are young, vulnerable, cute, and thus, particularly endearing. Given that people are especially susceptible to signals that indicate helplessness and extreme youth, it is to be expected that they would have also evolved to send signals that mimic those same traits.

On the other hand, parents, too, have psychological techniques at their disposal. Because of their greater age and experience, they often have something worthwhile to transmit to their offspring. Offspring, in turn, are therefore likely to be vulnerable to parents who exaggerate their wisdom (read here, manipulativeness), just as parents are vulnerable to offspring who exaggerate their neediness, or even, their competence. As Trivers also notes, with no small irony, it may therefore be significant that the traditional view of parent-offspring relations, which assumes that "father and mother know best," and which has never taken the perspective of offspring very seriously, is one that has been promulgated by adults!

Similarly, we can expect that parents would be likely to present their teachings, manipulations and arm-twisting as "for your own good," or even accompanied by protestations that "this hurts me more than it hurts you," in short, to emphasize the value of their opinions and advice, as well as the degree to which the parental perspective is solely in the best interest of the child.

New Perspectives on Child Development

Of all the new insights provided by a gene's-eye view of evolution, it may be that parent-offspring conflict has been the least explored. It has barely been acknowledged by child psychologists, not at all by psychoanalysts. Yet the possibilities are fascinating.

In a sense, we are all caught between the devil of altruism and the deep blue sea of selfishness. If we behave altruistically, we help others but at the risk of harming ourselves. If selfishly, we may do all right personally, but are likely to suffer various degrees of guilt. Indeed, perhaps the phenomenon of conscience (or the Freudian construct known as the "super-ego") represents a kind of parental victory, whereby parents induce their offspring to act in their—that is, the parents'—best interests, rather than the child's.[14]

For another example, take infantile (or rather, childhood) sexuality, a mainstay of psychoanalytic theory, yet poorly supported by facts. It, like the human conscience, may owe its existence to parent-offspring conflict. Here's how: when young children display a kind of intense (erotic?) interest in the opposite-sex parent, this could result in greater inclination on the part of that parent to invest in them. Not because of a direct, reciprocal sexual urge by the parent in question, but because it is only "human nature" to be flattered, to respond to such interest and attention by providing additional love, attention, resources, and so on. If so, then selection could favor such actions by children as part of their armamentarium in parent-offspring conflict, a way of counteracting what might otherwise be a parental inclination to discontinue investment, or at least, to provide less than the child might prefer. At the same time, it would be in the interest of the child not to be too overt about his or her approaches to the opposite-sexed parent, if only because this might arouse the ire of the same-sexed parent.

If in fact fathers are attached with special intensity to daughters, and mothers to sons (which is commonly believed, but not necessarily true), the above could be a partial explanation. From an evolutionary perspective, it is unlikely that children actually harbor desires for sexual union with their parents, and far more plausible that children occasionally respond provocatively to their opposite-sex parent in order to garner additional investment from them, as one of their tactical moves in the ongoing conflict between parent and offspring.

Another example, this one for the child psychologists: the phenomenon of the "terrible twos." It is a commonplace that toddlers go through a particular phase during which their behavior is notoriously trying. Is it merely coincidence that this happens at the same time that the parents are likely to be thinking of having another child? What more effective way for a child to reduce the likelihood of having a sibling, sooner perhaps than it might wish, than by announcing, "I'm already a handful; are you sure you're ready to deal with another one quite so soon?"

It may be noteworthy that the more an infant "wins" a weaning conflict the more likely he or she is to delay ovulation on the part of the mother, and therefore, the less likely it is that the mother will reproduce again. Given worldwide problems of overpopulation and under-nutrition, prolonged lactation would in fact seem to be a doubly good idea, and reason to root for the infant in this particular tug-of-war.

There is also something that may be even more terrible than the "terrible twos," at least from the parents' perspective. It is known as "copulatory interference." The title of the scientific paper reporting the phenomenon pretty much tells it all: "Responses of chimpanzees to copulation, with special reference to interference by immature individuals."[15] It turns out that among free-living chimpanzees at Jane Goodall's renowned research site in Tanzania, offspring are less than enthusiastic when their mother tries to copulate. The infant typically interposes him or herself between the mother and the sexually aroused adult male—a potentially risky undertaking—making noises as well as initiating physical contact with both male and female. During a sixteen-month period, 341 such instances were recorded.

Nor is this unique to chimpanzees: similar behavior is common among other primates, including macaques, patas monkeys, vervets and baboons, and is especially frequent when the mother is just beginning to resume sexual activity, around the time that the infant is being weaned. To be sure, copulatory interference could possibly be due to other causes—anxiety that the mother is being hurt, a voyeuristic interest in adult sexual behavior, and so forth—but it also makes perfect sense in terms of parent-offspring conflict, if such behavior can effectively delay the arrival of a sibling until the infant has received all the parental investment that it can get.

One of Freud's more celebrated cases was known as the "Wolf Man," not because the patient turned lupine during full moons, but because when very young he had apparently witnessed his parents having sexual intercourse, in the "doggie" position. Such "primal scenes" have long been considered emotionally shattering for young observers. There may be some truth to this, although it seems more likely that a toddler would find such occasions confusing than traumatizing. The real trauma for an inadvertently peeping-Tom child might be that parental intercourse signals the possible impending arrival of a competitor. Hence, we might expect an understandable ambivalence— on the child's part—about whether mom and dad are doing the right thing.[16]

People, unlike chimpanzees, are sometimes fortunate enough to leave their young children with a baby-sitter and get away for a week-end. Or at least they can lock the door.

Yet another unnerving possibility has turned up: Maybe parent-offspring conflict begins even earlier, *before* birth. Thus, some of the complications that bedevil pregnancy may be due to the fact that mother and fetus are not reading from the same genetic script.[17] Although to be sure, pregnancy is a cooperative endeavor—the mother is seeking to reproduce, and the fetus, to be produced—the fetus can be expected to seek more maternal resources (oxygen and nutrients) than the mother will be inclined to provide.

High blood pressure and pre-eclampsia, for example—two frequent complications of pregnancy, suffered by the mother—are mostly caused by fetal hormones that, in a sense, attempt to increase blood flow to the unborn child at the expense of blood flow to the mother's body. Blood sugar is another possible battleground. During the last trimester of pregnancy, the placenta (a product of the fetus, not the mother), secretes a hormone that interferes with the effect of the mother's insulin. The mother, at the same time, is producing more and more insulin. Why this particular tug-of-war? Insulin suppresses blood sugar levels, so by making insulin, the mother can be seen as attempting to restrict the amount of sugar—her nutrient—that is made available to the fetus, who in turn responds by trying to get as much as possible.

In most cases, things work out all right. But sometimes, the fetus "wins" and the result is gestational diabetes. (Is it a coincidence, by the way, that the gene responsible for producing the fetus's anti-insulin hormone, human placental lactogen, is provided by the *father*?)

This is admittedly speculative. But it is noteworthy that even the womb, seemingly sacrosanct site of unquestioned altruism and cooperation, is now undergoing scrutiny, yet another example of the startling new way of thinking that has been inspired by revolutionary, gene-centered biology.

Notes

1. R. L. Trivers. 1974. Parent-offspring conflict. American Zoologist 14: 249-264
2. T. C. Schneirla, J. S. Rosenblatt, and E. Tobach. 1963. Maternal behavior in the cat. In H. L. Reingold, ed. Maternal Behavior in Mammals, John Wiley & Sons: New York, 122-168.
3. The concept of parental investment was also developed by the mega-creative Robert Trivers.
4. Robert A. Hinde. 1977. Mother-infant separation and the nature of inter-individual relationships: Experiments with rhesus monkeys. Proceedings of the Royal Society of London, B. 196: 29-50.
5. N. B. Davies. 1976. Parental care and the transition to independent feeding in the young spotted fly-catcher (*Muscicapa striata*). Behaviour 59: 280-295.
6. Don't be misled by the thought that eventually, when Suzie reproduces, her offspring will be related to her by 50 percent, which is the same as the relatedness of Jimmy (Suzie's brother) to herself, and that therefore, Suzie should value Jimmy equally with herself. The flaw in such thinking is that Suzie and Jimmy, being of the same generation, are in a sense on a par. The relevant comparison is thus Suzie:Suzie = 100 percent, compared with Suzie:Jimmy = 50 percent; or, if we move ahead one generation, Suzie:Suzie's child = 50 percent, compared with Suzie:Jimmy's child = 25 percent. Any way you slice it, Suzie is twice as interested in herself or her lineage as she is in Jimmy or his.
7. P. W. Turke. 1988. Helpers at the nest: childcare on Ifaluk. In L. Betzig, M Borgerhoff Mulder and P. Turke, eds. Human Reproductive Behaviour—A Darwinian Perspective. Cambridge University Press: Cambridge, UK.

8. H. D. Schwbl, D. W. Mock, and J. A. Gieg. 1997. A hormonal mechanism for parental favouritism. Nature 386: 231.

9. Douglas W. Mock and Geoffrey A. Parker. 1997. The Evolution of Sibling Rivalry. Oxford University Press: New York.

10. M. A. Rodriques-Girones. 1996. Siblicide: The evolutionary blackmail. American Naturalist 148: 101-122.

11. F. Bryant Furlow. 1997. Human neonatal cry quality as an honest signal of fitness. Evolution and Human Behavior 18: 175-193.

12. G. B. Schaller. 1964. Breeding behavior of the white pelican at Yellowstone Lake, Wyoming. Condor. 66: 3-23; Chapman, F. M. 1908. Camps and Cruises of an Ornithologist. Appleton: New York.

13. van Lawick-Goodall, J. 1968. The behyaviour of free-living chimpanzees in the Gombe Stream Reserve. Animal Behaviour Monographs 1: 161-311; Yerkes, R. M. 1943. Chimpanzees, a Laboratory Colony. Yale University Press: New Haven, CT.

14. E. Voland and R. Voland. 1996. Parent-offspring conflict, the extended phenotype, and the evolution of conscience. Journal of Social and Biological Systems 18: 397-412.

15. C. Tutin. 1979. Animal Behaviour 27: 845-854.

16. Bear in mind that children need not consciously make this connection, any more than they need to understand the physiology of digestion in order to get hungry when they need nutrition.

17. David Haig. 1993. Genetic conflicts in human pregnancy. Quarterly Review of Biology 68: 495-531.

7

To Whatever Abyss

How satisfying it would be to pause now and summarize the wisdom of gene-centered revolutionary biology, to wrap everything up in a tidy package and say, "This is what it is to be made of genes, by genes and for genes, also to share genes with and interact with others, and at the same time to be an individual of the species *Homo sapiens*—in short, *this* is what it is to be human." But something is missing. And no one knows exactly what!

Referring to Shakespeare, T. S. Eliot once wrote that the best one could hope for was to be wrong about him in a new way. Shakespeare is that big, and that elusive. Human nature is also big, and elusive, even more so than Shakespeare. So maybe the best we can hope for is that gene-centered evolutionary biology will be wrong about human nature—what it means to be human—in a new way.

It is sometimes claimed that evolution has a big problem: it tries to do too much, and in the process, doesn't do anything. Evolution may thus appear to be like string theory or general relativity, a "theory of everything." Well, it isn't. Some of the most interesting aspects of revolutionary biology have to with its limitations and exceptions. And never are these limitations and exceptions more flagrant than when it comes to human beings. If people are, as Alexander Pope put it, "sole judge of all things, in endless error hurled, the glory, jest and riddle of the world," it is because *Homo sapiens* includes not only its nasty "animal" aspects of violence and selfishness but also those positive, equally animal components of benevolence and even self-sacrificing altruism.

In addition, there remains what Carl Sandburg calls, simply, "something else." And here may be the greatest challenge to evolutionary insights: not only the "something else" that leads to the writing of poems or the working out of great mathematical insights, but even the "something else" of dealing with each other in ways that would seem illogical and incoherent to a sensible animal.

Let's say you are home alone one rainy Saturday morning with your thirteen-year-old son. The day's schedule is well known to you both: in the morning, house chores, and in the afternoon, you take him to his soccer game. The evening is planned as well: he will have a friend over for the night, and you plan to watch old movies and eat popcorn. Sounds peaceful enough. At 11 o'clock that morning, however, you notice that his room is still a mess, the laundry isn't done, and he's talking on the telephone. You say, with some irritation, "Son, we have to leave in an hour and you haven't started your chores." He glares at you with apparent hatred, tells his friend that Mom (or Dad) is on the rampage again, and slams down the phone.

"Mom" (or "Dad") he says, "if you keep nagging me and butting in on my private phone calls, I'll get so mad that I'm going to destroy my computer!" You stare at him in total astonishment. What has gone wrong here?

First, there has obviously been a breakdown in benevolent reciprocity. You have planned to give quite a lot to your boy over the course of the day, and all you are asking in return is help with household responsibilities. You expected an hour or two of his time, doing something he didn't really want to do, in exchange for three to four hours of your time (as chauffeur) in addition to the extra trouble of having a household guest. The quid pro quo seemed fair and equitable to you, even weighted by kin selection criteria; namely, it will enhance your fitness to have your son the soccer player enhance his (even if you wouldn't have verbalized it quite this way). You are willing to give up a few extra hours of your day because he's your child and you want him to be happy, and to be outside, working his muscles and learning teamwork. You feel insulted, because what seemed like generosity on your part is being disdained by your son.

To top it off, there is something crazy about his threat. If you don't do as he wishes, he will destroy *his own personal computer!* The logic is disconnected. If you don't submit, he is saying, he will hurt himself to punish you; he will take an action to hurt you that will have far worse consequences for him, but he will do it anyway.

Biologists call this behavior spite, and no self-respecting honeybee, wolf, or baboon ever does it. Spite is defined as an action that hurts you more than your opponent, but you do it anyway. The classic formulation is "cutting off your nose to spite your face," that is, subjecting yourself to pain and torment. Why? To make yourself ugly! That is, for no good reason at all. There are no examples of nonhuman animals engaging in spite. Only people. Only people are such poor mathematicians, or so far out of touch with reality, or just so ornery, that they do something that is manifestly self-destructive in order to inflict pain on themselves or others, with no apparent return.

The teenager's case may be relatively easy to explain, attributable perhaps to an unfortunate combination of hormonal overload and intellectual under-functioning. But such behavior is not limited to frenzied adolescence. (Moreover, most animals have heavy doses of hormones, at least in season, and cerebrums that are probably dwarfed by even the most unreflective teenager.) An interesting clinical example of spite is found in one woman's deliberate response to her husband's adultery: she decided (quite consciously) to make herself ugly. Her husband was contrite about his affair; he apologized profusely and promised fidelity for the rest of his days. But the wife gave away all her attractive clothing, gained 100 pounds, quit cutting her toenails and fingernails, and stopped brushing her hair. This unfortunate couple lived another twenty years together, with the wife's visible poor health and poor self-care a constant and undeniable punishment to the husband, but worse yet for herself.

And here is a famous apocryphal story, but one that is, in some ways, painfully "real." It seems that a scorpion, seeking to cross the Jordan River, encounters a frog that is reluctant to carry him across, fearing the scorpion's sting. Finally, the frog is persuaded when the scorpion points out that if he were to sting the frog, both scorpion and frog would die. But sure enough, part way across the river, the scorpion can restrain himself no longer; he fatally stings the frog, who croaks out the agonized question, "Why?" just as he dies. To this, the scorpion answers, as he, too, slips beneath the waves: "Because this is the Middle East."

But "it" isn't just the Middle East (although spite is all too well represented in that part of the world). Perhaps spiteful behavior, for all its illogicality, is deeply *human* behavior, precisely because it is so illogical, and as such, part of what makes people human: the ability and inclination to act *contrary* to self-interest! (Unlike animals, whose actions are more predictable, more readily calculated and understood.) This was true for the great Russian novelist Fyodor Dostoyevsky, whose *Notes from the Underground* tells the story of a man whose driving motivation is spite: illogical, self-defeating, hurtful of others, the Underground Man is Dostoyevsky's answer to what he disdainfully called the Crystal Palace of Rationality, within which human beings are expected to behave in precise response to careful cost/benefit computations of precisely the sort that evolutionary biologists are inclined to make.

And so, Dostoyevsky's Underground Man intentionally subjects himself to humiliating circumstances, bringing pain on himself, his friends, his girlfriend, to prove that he is not a robot motivated by selfishness or altruism, or indeed, by anything that can be logically calculated and rationally understood. The Underground Man defiantly proclaims his humanity by behaving in a way that seems—to many of us—not only illogical but also inhumane.

This book is an enthusiastic endorsement of what I have called "revolutionary biology," the new gene's-eye perspective on life in general and human beings in particular. I remain encouraged by its growing concordance of theory and fact, and I stand by its legitimacy. At the same time, I acknowledge that *Homo sapiens* is a complicated creature, perhaps too complicated to be "explained" by any single biological theory, even one as encompassing as evolution.

It appears, for example, that there are occasions when human beings defy the biology that whispers within them, behaving with saintly charity and almost perfect altruism. As the millennium closes, it seems that there are even more occasions when the pendulum swings far the other way, when equally "non-biological" evil triumphs, and people behave with ultimately self-destructive sadism and spite. Perhaps unique among living things, human beings can ignore the cold calculus of self-interest and kin selection, and emerge as saintly or villainous, or just plain stubborn and incomprehensible. What to make of this?

Here is another way to look at it, admittedly a humorous fiction, but one that speaks to the unpredictable consequences of being a "B-cubed": a Beast with a Big Brain.

The Confusing Face of Duty

A huge octopus emerges from the ocean, wraps an oversized tentacle around the waist of a young woman, and proceeds to drag her into the sea. This memorable episode from Thomas Pynchon's vast, surreal novel, *Gravity's Rainbow*, has a happy ending, however, owing to the intervention of Mr. Tyrone Slothrop. First, he unavailingly beats the molluscan monster over the head with an empty wine bottle; then, in a stroke of zoologically informed genius, our hero offers the briny behemoth something even more alluring than a fair maiden: a delicious crab. It works, suggesting that this particular octopus conforms, at least in its dietary preference, to the norm for its species. Nonetheless, we learn that "In their brief time together, Slothrop formed the impression that this octopus was not in good mental health."

It isn't at all clear where the creature's mental derangement lies. There is no evidence, for example, that the octopus in question was spiteful; indeed, it behaved with a reasonable degree of healthy, enlightened self-interest in seeking first to consume the young lady, and then forgoing her for the even more delectable crab. Nonetheless, nature writer David Quammen may have been onto something when he pointed out that octopi generally—and not just Thomas Pynchon's admittedly fictional creation—might be especially vulnerable to mental disequilibrium, if only because one of their distinguishing characteristics is having immense brains. Mental strain is probably not unknown among animals, but there seems little doubt that it is especially well-developed in the species *Homo sapiens*, in which, as we have seen, the brain is especially large, in part perhaps because of the peculiar pressures of keeping a very complex social life in adaptive equilibrium.

As difficult as it must be for any animal to balance its various conflicting demands (to eat or sleep, to attack or retreat, etc.) such demands are probably greatest in the domain of social life. As confusing and stressful as it must be to predict the vagaries of weather, for example, the vagaries of one's fellow creatures have to be even more complex, confusing, and stressful. And when it comes to having a complicated and difficult social life, human beings are in a class by themselves. Moreover, they have been equipped by natural selection with a remarkably over-sized brain, one that does not satisfy itself with simply meeting the contingencies of daily life. Human neurons are obsessed with confronting all sorts of difficult issues, mostly of their own making. Small wonder that so many people, like Pynchon's octopus, are not in good mental health.

Some of the most anguished dilemmas encountered by the human "B-cubed" derive from conflicting loyalties, most of which can be illuminated by gene-centered evolutionary biology: conflicts between individual and gene, between what we want to do and what we feel that we should do, between cross-cutting obligations toward ourselves on the one hand and our family on the other, toward one family member versus another, toward friends versus the larger community, toward the law, the nation, other living things, the planet. Mark Twain suggested that human beings were the only living things that blushed—or had reason to. On a similar track, journalist Henry Hazlitt wrote that "Man is the only animal that laughs and weeps, for he is the only animal that is struck with the difference between what things are, and what they ought to be."

On the other have, there have been serious thinkers who maintained that for all the anguish of being human, deep inside, we all know full well what we ought or ought not to do. Immanuel Kant, widely acknowledged one of the greatest philosophers of all time, was convinced that human beings possessed an innate ability to understand and follow what he called the "categorical imperative" of ethical conduct.[1] For Kant, human inclinations often conflict with an appropriate sense of moral duty, and when this occurs, we nonetheless recognize the primacy of the latter, which typically involves hefty doses of self-denying altruism.

While the great philosopher Kant admired what he saw as this deep-seated human sense of moral duty, the not-quite-so-great rhymester Ogden Nash took a more down-to-earth view:

Oh Duty, Why hast thou not the visage

Of a sweetey or a cutey?[2]

Today's revolutionary biology helps us understand something—although certainly not everything—about why human duty is what it is, and why it appears as it does: sometimes sweet and cute, sometimes troublesome, even agonizing, but nearly always complicated and many-faceted. Unlike Thomas Pynchon's giant octopus, we need something more than a juicy and distracting morsel in order to avoid inflicting pain on others. We need to refrain from morally repulsive excesses of selfishness, from overdoses of self-destructive altruism, as well as from self-defeating spite. Our brains have been produced by self-serving genes, but these same self-serving inclinations have resulted in behavior that is often cooperative, social, and—at the level of bodies if not genes—altruistic. At the same time, we confront ourselves in societies that are, at best, uneasy compromises among the competing selfish tendencies of its component parts.

Admittedly, maybe the human species brings many of its difficulties upon itself, not the least because it is so excessively cerebral (even, as in the case of saintliness and spite, clever enough to outsmart its own genes). But the conflicting pressures of selfishness and altruism are so difficult to unravel that perhaps our species can only restore its mental health by employing that wonderful brain to reflect upon its own evolutionary situation.

Ethical Dilemmas

Just as nature is said to abhor a vacuum, it abhors true altruism. Society, on the other hand, adores it. Most ethical systems advocate undiscriminating altruism: "Turn the other cheek," we are told. "Virtue," we are advised, "is its own reward." Such sentiments are immensely attractive, not only because they are how we would like *other* people to behave, but also because at some level, we wish that we could do the same. It is also possible that "society" exists as we know it precisely because of our large brains, and the previously alluded to capacity for delayed gratification operating for the "good of the group."

On the other hand, as philosopher David Hume wrote in his *Enquiry Concerning Human Understanding*, "It is not irrational for me to prefer the destruction of half the world to the pricking of my finger." More than 200 years ago, people were made uncomfortable by such sentiments, and they still are. Biology helps us understand why: why culture is so opposed to selfishness and so fond of real altruism, and yet, why people are, at heart, so inclined toward the former and resistant to the latter.

Here is Hume, again:

Should a traveller give an account of men who were entirely divested of avarice, ambition, or revenge, who knew no pleasure but friendship, generosity, and public spirit, we should immediately detect the falsehood and prove him a liar with the same certitude as if he had stuffed his narration with centaurs and dragons.[3]

Such views are often put down as mere cynicism, and dangerous to boot. People generally yearn to be friendly, generous, and public-spirited, in short, altruistic, which is to say, good. As Bertolt Brecht put it in *The Threepenny Opera*, "we crave to be more kindly than we are." Throughout the world, not surprisingly, religious and moral systems teach the desirability as well as the possibility of striving to be good, even if not downright saintly.

One of the oldest debates among philosophers, ethicists, and theologians concerns this fundamental division of human nature: Are people naturally generous, altruistic, group-oriented, pro-social; that is, basically good? Or are we nasty, selfish, always looking out for private gain; that is, basically bad? Jean-Jacques Rousseau, with his idealization of the noble savage, "born free and everywhere in chains," embodied the former view. Thomas Hobbes represented the latter, dourly warning that natural life was "solitary, poor, nasty, brutish and short," culminating in a "warre of each against each."

This is more than a dry argument among professional scholars. Important things are at stake. Thus, if Hobbes is right and people are fundamentally selfish and antisocial, justification exists for police, jails, stern discipline, and large military establishments. If people are inherently evolved to be "bad," then social systems based on punishment or deterrence are likely to be more effective than any based on positive incentives, encouragement or social rewards. The argument might go like this: people are inherently self-serving and greedy, even potentially violent. The only way to make life tolerable among such nasty creatures might therefore be to make socially undesirable behavior so costly that such instincts would be inhibited. Rather than trust to the "better angels of our nature," lets acknowledge that human nature is composed more of devils than of angels, and be prepared to keep those devils at bay (or if need be, in jail).

Take shoplifting. It's hard to go to a store and see pretty things that one can't afford. It is therefore tempting to steal, unless the price—combined with the probability—of being caught is greater than the benefit of acquiring the article. Thus, harsh penalties for shoplifting inhibit or deter the "natural" impulse to steal. Spend your resources, therefore, on jails, and courts. As to criminals, lock 'em up and throw away the key. Don't even try to provide much in the way of education, job opportunities or medical care for the poor, since they're likely to remain a rotten, scurvy lot in any event. Instead, maintain a large and ferocious police force to keep them in line. And while you're at it, invest heavily in another kind of "police force" to be used in keeping the malefactors from other countries in line, too: the military.

If, on the other hand, Rousseau is right and people are good, noble, and well-meaning at heart, then there is far less reason for punitive measures at home or for war-fighting abroad. If people are basically good, then they are educable, and a more optimistic case can be made for social intervention at the public level.

Of course, real life isn't so simple. If people really are "inherently" selfish and nasty, perhaps this is all the *more* reason for programs to train or encourage them to be otherwise. And if people are good at heart, then maybe the best thing is to leave them alone, and let that goodness shine through.

Things aren't so simple from an evolutionary viewpoint, either. Thus, insofar as this book's genetic perspective is accurate, genes are, in fact, selfish. But this selfishness is often achieved by an array of altruistic, pro-social acts: toward relatives, reciprocators, potential reciprocators, even the larger group. It is a matter of seeing the glass of human selfishness as either half full or half empty.

Marxists are followers of Rousseau at heart, seeking to base an economic system on the idea that people are so fundamentally benevolent and inclined to share that they would willingly participate in a regime that took "from each according to his ability," and that gave "to each according to his need." In the realm of nation-states, capitalistic selfishness has proven to be more effective in motivating private effort and economic growth, just as in the realm of evolutionary biology, genuine altruism is rare and likely to occur, if at all, only under duress or manipulation.

But at the same time, conservatives—arguing for a Hobbesian view of human nature—are left equally unsupported. The Hobbesian quest for "power after power which endeth only in death" fails to reckon on kin selection, or reciprocal altruism. As we have seen, human beings are selfish and interested in themselves insofar as this enhances the success of genes that comprise them. They are also group-oriented or altruistic, insofar as this might enhance the success of genes identical to themselves, residing in the bodies of others. With remarkable consistency, actions are considered "moral" when they are in some sense altruistic, especially if that "altruism" extends beyond the immediate advantage of one's self or children.

The evolutionary perspective presented in this book thus produces an oddly ambivalent answer to the question: What is human nature? It suggests that whereas the genes that make up every human being are fundamentally selfish, expressing this selfishness can result in a kind of group-oriented altruism. Kin selection, for example, produces a powerful bias toward family members, including, perhaps, others who are psychologically identified as kin, even though they aren't. The urge for reciprocity generates another powerful current, flowing toward the exchange of favors and moral obligations. Reproduction and parental behavior leads to remarkable levels of self-sacrifice and devotion. Yet, genetic selfishness underlies it all. Alexander Pope concluded, with some satisfaction,

REASON and PASSION answer one great aim
That here SELF-LOVE and SOCIAL are the same.

And yet, even behavioral tendencies that are generally regarded as moral and desirable can fade imperceptibly into actions that are immoral and undesirable: excessive concern with one's inclusive fitness—to the detriment of others who aren't related—results in complaints of nepotism, something generally seen as unattractive, unfair, often illegal. Too much pro-social identification with the group can breed not only patriotism, but chauvinism, jingoism, bigotry, and warmongering. Reciprocity can lead to cheating, and parenting, to parent-offspring conflict.

At a level that is perhaps even more basic, there is the never-ending contention between selfishness and social obligation, what we might call "Maggie's Dilemma," after the beautiful heroine of George Eliot's novel, *The Mill on the Floss*. Maggie could become the wife of either of two attractive young men (the selfish, personally fulfilling route), but at the cost of mortifying her family, especially her rigid and disapproving brother. Or she could deny both prospective husbands (and thus, herself), while remaining true to her social obligations. Maggie's Dilemma is stated by Eliot as follows: "The great problem of the shifting relation between passion [selfishness] and duty [social altruism] is clear to no man . . . " Maggie resolves it in favor of the latter: "I cannot take a good for myself that has been wrung out of their [her family's] misery."

For most of us, Maggie's Dilemma remains very real. Gratify yourself, or your family? Cheat a little here and there, or be an upstanding, honest person? Discard your trash, or recycle it? Be bad, and satisfy your "passion" or be good, and do your "duty"?

No one said these issues would be easy. As noted earlier, perhaps this is one reason why human beings often find their mental health under assault. In one *Sesame Street* song, Kermit the Frog points out "It's not easy being green." It's not easy being human, either.

Take the remarkable opening pages of a recent novel, *Enduring Love*, by Ian McEwan.[4] Joe is enjoying a picnic in the English countryside when he hears a shout for help and discovers a man struggling with a large gas balloon, being tossed about by the wind. There is a little boy in the basket. Joe and four other men grab the balloon by a trailing rope, but, just when it seems that they are going to rescue the boy, a sudden powerful gust of wind carries the balloon and its occupant over the edge of an impossibly steep slope. Joe and three of the other men let go immediately; the fourth holds on, but not for long. He falls to his death, having tried to save the boy (who, ironically, manages to survive uninjured).

As Joe reflects on the event—and how he and the three other men had released the rope, choosing to save themselves rather than the child—he acknowledges its primordial quality:

> This is our mammalian conflict, what to give to others and what to keep for yourself. Treading that line, keeping the others in check and being kept in check by them, is what we call morality. Hanging a few feet above the Chilterns escarpment, our crew enacted morality's ancient dilemma: us, or me.

In this case, Joe didn't know the boy in the balloon, and certainly wasn't related to him. Therefore, "us" didn't outweigh "me." On the other hand, the same was true for the man who died, yet he didn't let go, until it was too late. Maybe he was following a different, "higher" morality (lethally elevated, a cynic might add!). Or maybe he just held on too long, then couldn't let go safely even if—when?—he wanted to.

Part of the difficulty of being human is the often agonizing need to decide on one's own ethical precepts, to establish the boundaries of what is desirable, the limits of what is acceptable, and when, and why. Can evolutionary psychology—or, as I have called it, revolutionary biology, or a gene's-eye view of evolution—help? I'm not sure. It might even hurt, if people take selfishness as its primary lesson (and especially if they go further and commit the "naturalistic fallacy" and assume that anything that is "natural" is necessarily good.) On the other hand, if people focus instead on the altruistic side of the evolutionary coin—on the prosocial, reciprocating, kin- and group-oriented aspect of human nature—they are likely to derive a very different take-home message. Moreover, gene-centered evolutionary thinking might serve to expand the sense of self and emphasize interrelatedness: Altruism aside, just consider all those genes for cellular metabolism, for neurotransmitters and basic body plans, all of them shared with every living thing, competing and pushing and somehow working things out on a small and increasingly crowded planet. There, by the grace of evolution, go a large part of "ourselves."

Society's Perspective

Life nonetheless remains difficult, not only for the individual, but for society, too. It is not easy to deal with so confused, confusing, and obstreperous a creature as *Homo*, ostensibly *sapiens*. From the viewpoint of the larger group, it is people's selfish tendencies—selfish at the level of bodies—that seem to pose the greatest threat, and have generated the most serious efforts at control.

If actions that are in one's self-interest are branded with pejorative titles such as selfish, antisocial, immoral, etc., then it may be easier to prevent them. People are likely to stifle any inclinations labeled despicable. On the other hand, an individual's actions that are motivated by self-interest may be considered entrepreneurial, assertive, or creative. Someone with little self-motivation may be perceived of as lazy, passive or dependent. In other words, society condemns the extremes while rewarding those whose self-serving actions succeed, so long as they are carried out without obvious malevolence.

At the same time, we can expect that society will often call for real altruism, not because it is good for the altruist but because it benefits those who receive. (If it were clearly good for the altruist, then society wouldn't have to call for it! In fact, it is precisely because altruism is generally *not* good for the altruist that social pressures are so often focused on producing it.) The German philosopher-genius-madman-poet Friedrich Nietzsche was probably the most articulate spokesperson for the cynical view that society encourages self-sacrifice because the unselfish sucker is an asset to others:

> virtues (such as industriousness, obedience, chastity, piety, justness) are mostly *injurious* to their possessors. . . . If you possess a virtue . . . you are its *victim!* But that is precisely why your neighbor praises your virtue. Praise of the selfless,

sacrificing, virtuous . . . is in any event not a product of the spirit of selflessness! One's "neighbor" praises selflessness because *he derives advantage from it!*[5] [Italics in original]

If Nietzsche is correct, then there is probably a distressingly manipulative quality to morals, to most religious teachings, to the newspaper headlines that celebrate the hero who leaps into a raging river to rescue a drowning child, to local Good Citizenship Awards and PTA prizes.

"That man is good who does good to others," wrote the seventeenth-century French moralist Jean de la Bruyere. Nothing objectionable so far; indeed, it makes sense (especially for the "others."). But de la Bruyere goes on, revealing a wicked Nietzschean cynicism:

if he suffers on account of the good he does, he is very good; if he suffers at the hands of those to whom he has done good, then his goodness is so great that it could be enhanced only by greater suffering; and if he should die at their hands, his virtue can go no further; it is heroic, it is perfect.

Such "perfect" heroism can only be wished on one's worst enemies.

Exhortations to extreme selflessness are easy to parody, as not only unrealistic but also paradoxically self-serving. Yet, the more we learn about biology, the more sensible becomes the basic thrust of social ethics, precisely because (even with the meliorating effects of kin selection and reciprocity), nearly everyone, left to his or her devices, is likely to be selfish, probably more than is good for the rest of us. Philosopher and mathematician Bertrand Russell pointed out that "by the cultivation of large and generous desires . . . men can be brought to act more than they do at present in a manner that is consistent with the general happiness of mankind."[6] Society is therefore left with the responsibility to do a heck of a lot of cultivating.

Seen this way, a biologically appropriate wisdom begins to emerge from the various Commandments and moral injunctions, nearly all of which can at least be interpreted as trying to get people to behave "better," that is, to develop and then act upon large and generous desires, to strive to be more amiable, more altruistic, less competitive and less selfish than they might otherwise be.

The more we look at the situation, the more it becomes, as the King of Siam puts it in *The King and I*, "a puzzlement." Thus, although people are widely urged to be kind, moral, altruistic, and so forth—which suggests that they are basically *less* kind, moral, altruistic, etc., than is desired—it is also common to give at least lip service to the precept that people are fundamentally good.

It appears that there is a payoff in claiming—if not acting—as though most people are good at heart. "Each of us will be well advised, on some suitable occasion," wrote Freud, in *Civilization and Its Discontents,* "to make a low bow to the deeply moral nature of mankind; it will help us to be generally popular and much will be forgiven us for it." Why are people generally so unkind to those who criticize the human species as nasty and selfish? Maybe because of worry that such critics might be seeking to justify their own unpleasantness by pointing to a general unpleasantness on the part of others. And maybe also because most people like to think of themselves as benevolent and altruistic, or at least, to think that other people think of them this way. It seems likely that the cynic is harder to bamboozle.

In any event, Freud has generally been far more in the Hobbesian (people-are-selfish-and-nasty) camp than that of Rousseau (people-are-altruistic-and-good). In *Civilization and its Discontents*, perhaps his most pessimistic book, Freud went on to lament that one of education's sins is that

> it does not prepare them [children] for the aggressiveness of which they are destined to become the objects. In sending the young into life with such a false psychological orientation, education is behaving as though one were to equip people starting on a Polar expedition with summer clothing and maps of the Italian Lakes. In this it becomes evident that a certain misuse is being made of ethical demands. The strictness of those demands would not do so much harm if education were to say: "This is how men ought to be, in order to be happy and to make others happy; but you have to reckon on their not being like that." Instead of this the young are made to believe that everyone else fulfills those ethical demands—that is, that everyone else is virtuous. It is on this that the demand is based that the young, too, shall become virtuous.

On the other hand, the expectation that others will be aggressive and nasty can become a self-fulfilling prophecy, especially if it leads people to be aggressive and nasty first.

But hold on: The human paradox is even more complicated. In his book, *The Ghost in the Machine*, Arthur Koestler suggested that violence is not caused so much by an excess of individualistic selfishness as by too much group-centeredness . . . in a sense, an excess of altruism:

Selfishness is not the primary culprit . . . appeals to man's better nature are bound to be ineffectual because the main danger lies precisely in what we are wont to call his 'better nature.' . . . the crimes of violence committed for selfish, personal motives are historically insignificant compared to those committed . . . out of a self-sacrificial devotion to a flag, a leader, a religious faith or a political conviction.[7]

According to Koestler, we might temper our admiration for altruistic enthusiasm by considering the "downside" of such unselfishness:

The total identification of the individual with the group makes him unselfish in more than one sense. It makes him indifferent to danger and less sensitive to physical pain . . . . It makes him perform comradely, altruistic, heroic actions—to the point of self-sacrifice—and at the same time behave with ruthless cruelty towards the enemy or victim of the group. But the brutality displayed by the members of a fanatic crowd is impersonal and unselfish; it is exercised in the interest or the supposed interest of the whole; and it entails the readiness not only to kill but also to die in its name. In other words, the self-assertive behavior of the group is based on the self-transcending behavior of its members, which often entails sacrifice of personal interests and even of life in the interest of the group. To put it simply: the egotism of the group feeds on the altruism of its members.

By this point, it should be clear that in generating human beings, evolution has put together a strange amalgam of selfish nastiness and altruistic kindliness, sometimes fading into altruistic violence. Neither altruism nor selfishness is more "biological" than the other, and yet, the argument that human beings are naturally group-oriented, cooperative, or altruistic is widely seen as less "instinctivist" or "genetically determinist" than the alternative view: that we are all somewhat competitive, aggressive, and selfish. Biology does not have a monopoly on nasty, selfish behavior, nor does social learning work only on behalf of altruism. People can learn violence and they are just as much "naturally" altruistic as they are "naturally" selfish. Few would object to the suggestion that human beings have a biologically given tendency to find certain physical stimuli noxious: red-hot objects, cold water when we are sick or already shivering, poisonous insects or snakes, and so on. Then why shouldn't we also find certain social actions noxious or unpleasant, and others, user-friendly and compatible with our natures?

Self-Deception, Self-Knowledge, and the Cheshire Cat

Among those things likely to appeal to our evolutionary sweet-tooth, there are—in addition to nepotism, reciprocity, parental care, and so forth—some surprising tid-bits. As we have seen in the context of reciprocity, one of these may even be a fondness for self-deception.

Richard Alexander has emphasized that human beings may actually have been selected, naturally, to deny the evolutionary basis of their own motivations. There are two possible reasons for this. First, people in the past would probably have been at an advantage if they claimed to be altruistic rather than selfish. And second, they may have honestly believed that they were altruistic rather than selfish, because the most successful deceivers are those who believe their own deceptions. Most people cherish the notion that they are fully prepared to behave altruistically, and selflessly.

Both Robert Trivers and Richard Alexander have also called attention to what they see as this peculiar biological bar to self-knowledge: They argue that to admit evolutionary biologists are correct, that we are inclined to behave in accord with gene-based strategies designed to help ourselves (really, our genes), is to acknowledge that deep inside, we are nasty, selfish, egoistic brutes. It is to make public confession of what is, for most people, their deepest and best hidden secrets. Moreover, to admit the underlying importance of fitness maximization is—at least in the environments in which human beings evolved—likely to diminish that fitness. No wonder most people don't do it.

Most evolutionary biologists are agreed that the human brain—like everything else about the human body—has evolved for one purpose and one purpose only: as Edward O. Wilson put it, "the brain exists because it promotes the survival and multiplication of the genes that direct its assembly."[8] This is a secretly stunning idea. It says, for example, that, the brain has *not* evolved to provide human beings with an objectively accurate picture of the outside world, except insofar as such a picture might aid in promoting the ultimate goal of success for the genes involved.

Looking at living things, we see marvelous examples of how evolution has solved various problems posed by the natural world. Whales and sharks, for example, are only distantly related (the former are mammals, the latter, primitive fish). Yet they have converged on virtually identical ways of coping with the hydrodynamic properties of water. To these one can add submarines, designed by human beings but with comparable goals. In much the same way, living things have contrived to solve such problems as how to maximize their inclusive fitness (via parenting, nepotism, reciprocity, and so forth). But in all such cases, the ultimate goal is fitness, and the basic rule of thumb is: whatever works.

It matters not a whit that bottle-nosed dolphins or great blue sharks understand neither themselves nor the hydrodynamics of water, so long as they are constructed—and behave—appropriately. In cases of this sort, insight is strictly optional. Better yet, it is irrelevant. So is absolute accuracy: dolphins and sharks would generally be no less fit if they totally *mis*understood the ultimate reality of water, so long as their anatomy, physiology, and behavior did not betray their ignorance.

The urge for self-knowledge presents human beings with a challenge: to figure out what our genes are up to. And that challenge is magnified by the likelihood that when our genes put us together, self-knowledge was not high on their agenda. It probably was not there at all. In fact, our mental processes may even have been established with a particular bias *against* self-knowledge, if such insights threatened the evolutionary success of those possessing it.

Maybe this is why so much of people's mental life floats beneath the surface, like an iceberg. If people went about their lives flagrantly seeking to enhance their fitness—often at the cost of someone else's—they would be unlikely to make friends and influence people. More likely, they would be set upon, locked up, shunned, or even killed. This might explain the "adaptive significance" of the unconscious, a repository for—among other things—a ravening core of genetic selfishness known to Freudians as the "id." The unconscious may therefore have been positively selected for, not just a by-product of something else, or a historical vestige, but something of definite survival and reproductive value. Freud's "super-ego" would then consist of the collective advice, moralizing and urging of society and parents, seeking to make us more altruistic than we are inclined to be. And the "ego" is the integration of the two, the public face we wear to meet the faces that we meet.

Animals, by contrast, are widely thought to lack consciousness. But maybe the opposite is true, and animals are more likely to lack an *unconscious*. Although they, too, have occasional need to be deceptive and manipulative, such requirements pale in comparison to the misrepresentation that human beings require of themselves, on a regular basis. And so, it may be that whereas people have consciousness combined with a hefty, adaptive dose of the unconscious, animals are overwhelmingly conscious, and little else!

"Man is but a reed, the most feeble thing in nature," wrote Blaise Pascal, brilliant seventeenth-century French mathematician, philosopher, and religious mystic. In this, perhaps his most famous passage, however, Pascal added an important qualification:

> but he is a thinking reed. . . . If the universe were to crush him, man would still be more noble than that which killed him, because he knows that he dies and the advantage which the universe has over him; the universe knows nothing of this.

Substitute "evolution" or "natural selection" for "the universe," and you have a good statement of humanity's claim to glory. We, as a species and as individuals, are formed by evolution and subject to its strictures. We could not live as amoebas, honeybees, or as African lions. But unlike amoebas, honeybees, or lions, which can only be what their biology has narrowly constructed them to be, human beings—because of their large brains and flexible behavior—have the luxury (or the curse!) to step outside themselves, and look at their own nature, their own inclinations. They can inquire into the role of evolution in their lives, just as Pascal's humanity can glory in knowing its relationship to the universe, even if part of that relationship includes eventually being killed, and knowing it. In this view, it is when human beings confront their biological natures that they have a particular obligation to be thoughtful, and indeed, to be ethical as well as self-critical.

There is a widespread tendency to overrate the flexibility and capacities of human beings, presuming that our species might seem to have literally no limits. After all, the human species can fly more swiftly than birds, even faster than sound. It can communicate across thousands of miles, redirect the flow of rivers, tinker with subatomic particles, land on the moon. And so, some of us get impatient with the idea of any restrictions whatever. Such a species-wide *hubris* can easily lead to unwise attacks on the natural world and to unrealistic expectations of the human ability to get itself out of trouble, trouble of the sort that a bit of biologically informed humility might have prevented.

Every one of the world's formerly communist countries, for example, had slogans about the new Soviet (or Polish, or Bulgarian) man, or woman. In fact, neither a new man nor a new woman was ever created, in the sense of people who are fundamentally different from other men and women living in other socioeconomic systems. Sad to say, it is just not possible to make a dramatically—that is, biologically—new man or woman, whether communist or capitalist, or any other kind, for that matter. We are stuck with the old models. What we can do, however, is make *better* men and women, perhaps by helping them to be aware of their selfishness and their altruism; in short, their humanity.

Being a better human being may also require being willing to act in opposition to some deeper inclinations. By the end of the nineteenth century, Thomas Huxley was perhaps the most famous living biologist, renowned in the English-speaking world as "Darwin's bulldog," for his fierce and determined defense of natural selection. In 1893, Huxley gave a lecture titled "Evolution and Ethics," to a packed house in Oxford, England. "The ethical progress of society depends," he announced, "not on imitating the cosmic process, [that is, evolution by natural selection] still less in running away from it, but in combating it."[9] After all, evolution is, as we have seen, a profoundly selfish process, at least at the level of the gene. It is also entirely indifferent to ethics or morality; indeed, by its callous indifference, much of evolution is deservedly offensive to most human judgments of right and wrong.

It may seem impossible for human beings to "combat" evolution, since *Homo sapiens*—no less than every other species—is one of its products. At yet, Huxley's exhortation is not unrealistic. It seems likely, for example, that to some extent each of us undergoes a trajectory of decreasing selfishness and increasing altruism as we grow up, beginning with the infantile conviction that the world exists solely for our personal gratification and then, over time, experiencing the mellowing of increased wisdom and perspective as we become aware of the other lives around us, which are not all oriented toward ourselves. In her novel, *Middlemarch*, George Eliot noted that "we are all born in moral stupidity, taking the world as an udder with which to feed ourselves." Over time, this "moral stupidity" is replaced—in varying degrees— with ethical acuity, the sharpness of which can largely be judged by the amount of unselfish altruism that is generated.

Many sober, highly intelligent scientists and humanists misunderstand the connection between evolution and morality, grimly determined that evolutionary facts are dangerous because they justify human misbehavior. Jerome Kagan, renowned developmental psychologist, exemplifies this blind spot.[10] "Evolutionary arguments," he writes, "are used to cleanse greed, promiscuity, and the abuse of stepchildren of moral taint." Similar arguments have in fact been used in this way, in the unlamented days of social Darwinism, long before the advent of revolutionary biology. But no longer. Today, they are used to *understand* greed, promiscuity, and the abuse of stepchildren—and also to help understand parenting, nepotism, reciprocity, friendship, parent-offspring conflict, courtship, altruism, and bigotry, to name just a few.

Despite the role of biology in influencing behavior, human beings remain, after all, the most flexible, adaptable creatures on Earth. Think for a minute about the mundane behavior of toilet-training. All people, all over the planet, seem to have a hard time with it, a difficulty that is especially surprising in view of the fact that dogs and cats—creatures that are certainly stupider than *Homo sapiens*—can generally be house-broken in a few weeks or days. Why do children take literally years to learn something so simple?

The answer seems to lie in the evolutionary past, ours and theirs: Dogs and cats evolved in a two-dimensional world, in which it was adaptive to avoid fouling their dens. And so, they have developed a readiness to learn where—and where not—to relieve themselves. By contrast, human beings, like other primates, evolved in the trees. Tidy potty habits simply aren't an issue for arboreal creatures (although they can be a problem for those poor unfortunates down below!). This is almost certainly the reason why other primates, such as chimpanzees or monkeys, also resist toilet-training. It goes against a deep-seated inclination.

But is it hopeless? Not at all. The likelihood is that virtually everyone reading this book has been successfully toilet-trained, his or her genetic predispositions notwithstanding. The point is this: Whatever the specific genes that may ultimately be revealed to underlie human inclinations for altruism or selfishness, any primate that can learn to control its bowels has amply demonstrated that it can overcome powerful genetic tendencies, if it decides to do so.

Accordingly, even that most cherished human possession, free will, is not excluded by the genetic perspective espoused in this book. Indeed, it is precisely in biological terms that free will can best be understood. The extraordinary liberation of behavior from genes that characterizes human action is not a random happenstance, nor is it a sign of divine providence. Most likely, it, too, is a strategy upon which *Homo sapiens* genes have settled. Insofar as human beings and their genes experience a higher degree of reproductive success by giving their behavior a hefty dose of independence from otherwise suffocating genetic control, then independence it is. But, paradoxically, such independence is itself a result of selection for genes that permit precisely this independence and flexibility—that is, free will.

Human beings, more than any other living things, are characterized by an almost unlimited repertoire, a behavioral range that exceeds that of any other living creature. People are of the wilderness, with beasts inside, but, as Carl Sandburg put it, each human being is "the keeper of his zoo." The wide-ranging human repertoire is not evidence of a lack of evolutionary influence. Rather, it is a *result* of selection for being a good zoo-keeper, for having a flexible, wide-ranging repertoire.

"Gene-centred theories are often reviled," writes biologist David Haig,

> because of their perceived implications for human societies. But, even though genes may cajole, deceive, cheat, swindle or steal, all in pursuit of their own replication, this does not mean that people must be similarly self-interested. Organisms are collective entities (like firms, communes, unions, charities, teams) and the behaviours and decisions of collective bodies need not mirror those of their individual members. As I write this paragraph, my replicators . . . are in constant debate, even dissension, yet somehow I muddle through. I am glad that I am not a unit of selection.[11]

One goal in seeking to clarify the evolutionary biology that underpins humanity is not to identify our limits, the supposed boundaries of human potential, so as to live within them. Instead, it is to help free people from constraints, by identifying where any such impediments may lurk. Maybe even to make us glad.

But where is humanity's biological self-knowledge likely to lead? No one knows, but it is probably the obligation of *Homo sapiens* to follow. Maybe it will assist in some of humankind's oldest, yet most recent struggles, such as the unending confrontation with racism and bigotry, helping to mitigate Einstein's Lament (why politics is so much harder than physics), aiding everyday people as they seek to unravel Maggie's Dilemma (how to navigate between the Scylla of selfish desire and the Charybdis of social obligation), in mitigating—or at least, making sense of—the human penchant for in-group amity and out-group enmity, and, overall, assisting human beings in their never-ending quest to Know Themselves.

When Alice was lost in Wonderland, she asked the Cheshire Cat:

"Would you tell me, please, which way I ought to go from here?"
"That depends a good deal on where you want to get to," said the Cat.
"I don't much care where—" said Alice.
"Then it doesn't matter which way you go," said the Cat.
"—so long as I get *somewhere*," Alice added as an explanation.
"Oh, you're sure to do that," said the Cat, "if you only walk long enough."

Like the Cheshire Cat, biology cannot tell human beings where to go. But once decided, it may help us chart a course.

Let's conclude with this chapter's title, a phrase borrowed from physician/scientist Homer W. Smith, whose memorable book, *Man and His Gods*, traced the origin of religion. In it, Dr. Smith concluded that it is time for *Homo sapiens* to cast off fear and superstition, and follow the road of intelligent, skeptical, scientific inquiry wherever it might lead—to whatever abyss.

Notes

1. Kant formulated his ethic as follows: behave as though you would like your action to become a universal maxim for the behavior of others. In our terms, Kant's categorical imperative would favor selfishness if one could be satisfied that universal selfishness would lead to a better (or at least, a good) world. Alternatively, if universal altruism is preferable, then individual altruism is the imperative. Perceptive readers may have noted that this system neatly anticipates—and solves—the Prisoner's Dilemma.
2. Ogden Nash. 1959. Kind of an Ode to Duty. Verses from 1929 On. Random House: New York.
3. David Hume. 1772 (1985). Essays and Treatises on Several Subjects. Liberty: Indianapolis
4. 1997. Doubleday: New York.
5. R. J. Hollingdale, ed. 1977. A Nietzsche Reader, Penguin: Hammondsworth, 101.
6. Bertrand Russell. 1935. Religion and Science. Henry Holt & Co.: New York.
7. Arthur Koestler. 1967. The Ghost in the Machine. Pan Books: London.
8. E. O. Wilson. 1978. On Human Nature. Harvard University Press: Cambridge, MA.
9. Thomas Huxley. 1989. Evolution and Ethics. Princeton University Press: Princeton, NJ.
10. 1998. Three Seductive Ideas. Harvard University Press: Cambridge, MA.
11. David Haig. 1997. The Social Gene. In Behavioural Ecology, an evolutionary approach. J. R. Krebs and N. B. Davies, eds. Blackwell Science: London.

Index

(note: because certain terms and concepts, such as evolution, natural selection, altruism, selfishness, human nature, gene, kinship, kin selection, and "revolutionary biology" appear throughout this book, they are not specifically indexed here)